Phosphorus and Nitrogen Removal

From

Municipal Wastewater

Principles and Practice, Second Edition

 LEWIS PUBLISHERS

Library of Congress Cataloging in Publication Data

Phosphorus and nitrogen removal from municipal wastewater: principles and
practice/Richard I. Sedlak, editor.—2nd ed.
    p. cm.
  Includes bibliographical references.
  ISBN 0-87371-683-3.
    1. Sewage—Purification—Nitrogen removal. 2. Sewage—Purification—Phosphate
removal. 3. Sewage disposal—United States—Cases studies. I. Sedlak, Richard I.
  TD758.P46   1991
  658.3'57—dc20

91- 29578
CIP

Notice

The data and information contained in this manual were compiled by the contributing authors and any views or opinions expressed are solely those of the authors. The Soap and Detergent Association does not warrant, either expressly or impliedly, the accuracy or the completeness of the information contained in this manual, and it assumes no responsibility or liability for the use of the information. Further, nothing herein constitutes an endorsement of, or recommendation regarding any product or process by The Soap and Detergent Association.

EDITOR

Richard I. Sedlak Technical Director, The Soap and Detergent Association, New York, NY

CONTRIBUTORS

Yerachmiel Argaman Professor of Civil Engineering, Israel Institute of Technology, Haifa, Israel

James L. Barnard Meiring & Barnard, Sunnyside, South Africa

Glen T. Daigger Vice President and Assistant Director, Wastewater Reclamation Discipline Group, CH2M HILL, Denver, CO

W. Wesley Eckenfelder, Jr. Chairman Emeritus and Senior Technical Director, Eckenfelder, Inc., Nashville, TN

Slawomir W. Hermanowicz Assistant Professor of Sanitary Engineering, Department of Civil Engineering, University of California at Berkeley, CA

David Jenkins Professor of Environmental Engineering, Department of Civil Engineering, University of California at Berkeley, CA

Steven R. Polson Process Engineer, CH2M HILL, Denver, CO

Thomas W. Sigmund Process Engineer, CH2M HILL, Milwaukee, WI

H. David Stensel Professor and Director of the Environmental Engineering Program, University of Washington, Seattle, WA

Contents

Contents (continued)

Contents (continued)

Contents (continued)

6 Principles of Biological Phosphorus Removal 141

H. David Stensel

Contents (continued)

Figures

Figures (continued)

Figures (continued)

Tables

Chapter 1

Introduction

As the approach of removing phosphorus from municipal wastewaters to control nuisance aquatic plant growth enters its third decade of application in the United States, two significant trends can be observed. First, it is a well established approach to ameliorate water quality problems that will be increasingly applied over the next decade and beyond. For example, twenty percent of the total U.S. treatment capacity is expected to be capable of removing phosphorus in the year 2000, a percentage that is twice as great as in 1982(1).

Second, localized water quality problems can be expected to lead to lower and lower effluent phosphorus limitations. Historically, effluents limits of 1 or 2 mg total phosphorus (TP) per liter have been broadly applied in regions of the U.S.A., such as in the Great Lakes Drainage Basin (1 mg/L) and the Lower Susquehanna River Basin (2 mg/L). However, localized water quality conditions are leading to lower effluent phosphorus limits in some areas. One area where this has been the case for a number of years is the lower Potomac River Basin where municipal plants must meet discharge limits that are lower than 0.2 mg TP/L.

Concerns over nitrogen compounds have been primarily over ammonia toxicity to aquatic organisms, which has resulted in nitrification requirements being implemented more broadly than even phosphorus removal. Like for phosphorus, the extent of nitrification is expected to increase. It is anticipated that 27% of the total U.S.A. sewage flow will be treated for nitrification by the year 2000(1).

In contrast to phosphorus, requirements for nitrogen removal from municipal wastewaters historically have been applied on a limited basis in situations where nitrogen reductions are needed to correct localized water quality problems. However, an increasing trend in the future toward nitrogen removal requirements can be expected due to at least a couple of factors. First, nitrogen removal is now being considered on a broad-scale basis to reduce the availability of this nutrient to aquatic plants. For example, removal of nitrogen at municipal wastewater treatment plants is being considered throughout the Chesapeake Bay Drainage Basin. Second, broad-scale removal of nitrogen is being considered in areas where there is concern over the fate of nitrogen compounds in ground water drinking supplies that depend on recharge using municipal wastewaters. The proportion of the total U.S.A. sewage flow treated for removal of nitrogen is expected to double from 1982 levels by the year 2000, to approximately 2%(1).

As urban populations and, therefore, sewage flows increase and the accomplishments of current control programs become more apparent, consideration may be given in the future to more stringent municipal effluent limits for phosphorus and nitrogen to address local water quality problems.

1

All of the steps to date to control municipal phosphorus and nitrogen have not come without some considerable effort and cost. Nor will future reductions be effortless. However, programs to control nutrients over the past two decades have encouraged the development not only in the U.S.A. but elsewhere in the world of many treatment technologies for phosphorus and nitrogen removal. While dependable treatment technologies, such as chemical treatment for phosphorus removal, have been successfully utilized over these past two decades, improved understanding of the principles of the process has led to more efficient use of the approach. In addition, improved understanding of the mechanisms behind the biological removal of phosphorus will lead to broader and more efficient application of this approach. Similar comments can be made regarding the technologies for removal of nitrogen.

This document summarizes the available technologies for removing phosphorus and nitrogen from municipal sewage, with emphasis on those that are expected to see prominent use either because of their treatment capabilities or their ease and cost of operation, or both. The information is presented in two sequential blocks: one on the chemical, biological, and physical principles behind the available treatment technologies; a second on the design and operation of processes and systems based on these principles.

The information presented is based on available literature, as well as the experiences of the authors. It is presented in a format and with appropriate detail to assist those involved in the early stages of addressing the need to initiate nutrient removal at a facility or evaluating the feasibility of achieving lower effluent nutrient limits, such as personnel in government agencies, consulting and design engineers, and plant operators. Where appropriate the reader is directed to documents containing more detailed information on the design and operation of these types of facilities.

1.1 References

1.   Barth, E. F. Phosphorus control and nitrification processes for municipal wastewater. USA/USSR Bilateral Agreement on Water Pollution Control, 1985.

Chapter 2

Principles of Biological and Physical/Chemical

Nitrogen Removal

2.1 Introduction

Nitrogen exists in many forms because of the high number of oxidation states it can assume. In ammonium and organic nitrogen compounds, which are the forms most closely associated with plants and animals, its oxidation state is -3. At the other extreme, when nitrogen is in the nitrate form, its oxidation state is +5. In the environment, changes from one oxidation state to another can be accomplished biologically by living organisms. The most prevalent forms of nitrogen in wastewaters and, therefore, those which may require treatment, are organic, ammonium and nitrate nitrogen.

The presence of nitrogen in a wastewater discharge can be undesirable for several reasons: as free ammonia it is toxic to fish and many other aquatic organisms; as ammonium ion or ammonia it is an oxygen-consuming compound which will deplete the dissolved oxygen in receiving water; in all forms, nitrogen can be available as a nutrient to aquatic plants and consequently contribute to eutrophication; as the nitrate ion it is a potential public health hazard in water consumed by infants. Depending upon local circumstances, removal of all forms of nitrogen or just ammonium may be required. Both objectives can be achieved economically in biological treatment systems.

2.2 Sources of Nitrogen in Wastewater

Municipal wastewater of predominantly domestic origin contains nitrogen in the organic and ammonium forms. These are primarily waste products originating from protein metabolism in the human body. In fresh sewage about 60 percent of the nitrogen is in the organic form and 40 percent in the ammonium form. Bacterial decomposition of proteinaceous matter and hydrolysis of urea transform organic nitrogen to the ammonium form. Normally, very little (less than 1 percent) of the nitrogen in fresh sewage is in the oxidized form of nitrate or nitrite.

The average daily per capita production rate of nitrogen is approximately 16 grams. The nitrogen concentration in a wastewater depends on the per capita wastewater flow rate. Thus, for a flow rate ranging from 100 to 200 gallons per capita per day (gpcd), the calculated nitrogen concentration will range from 42 mg/L to 21 mg/L. Values reported in the literature (for U.S.A. cities) vary from 20 to 85 mg/L, as shown in Table 2-1.

3

Table 2-1. Nitrogen content of domestic sewage, mg N/L(1).

Nitrogen Form	Type of Sewage		
	Strong	Medium	Weak
Organic	35	15	8
Ammonia	50	25	12
Total	85	40	20

Industrial and commercial contributions, ground garbage and storm water will affect the nitrogen concentration in raw wastewater. In some treatment plants, nitrogen may also be introduced from recycle streams, such as supernatant from anaerobic digestors. In a typical municipal wastewater treatment plant, the soluble organic nitrogen (SON) remaining after biological treatment is in the order of 1 mg N/L.

2.3 Nitrogen Transformations in Biological Treatment Processes

The nitrogen transformations that may occur in biological treatment systems are illustrated in Figure 2-1. Systems can be designed and operated to influence this transformation scheme so as to achieve a desired effluent composition.

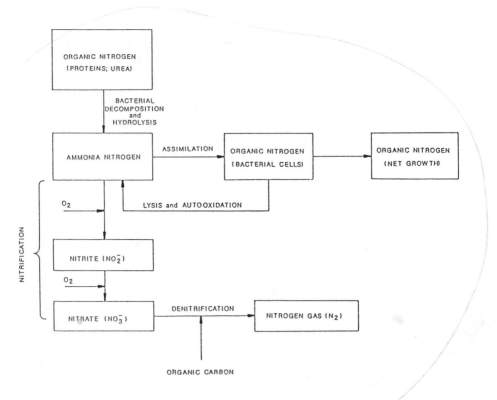

Figure 2-1. Nitrogen transformation in biological treatment processes.

As illustrated in Figure 2-1, organic nitrogen present in raw wastewaters may be transformed to ammonia through bacterial decomposition of proteinaceous matter and hydrolysis of urea. In any biological treatment system some bacterial growth always takes place. Since nitrogen constitutes 12 to 13 percent of cell dry mass, some ammonia nitrogen will be assimilated in newly formed cells. Depending on the treatment process and the loading condition, cell autoxidation and lysis will also occur. Hence part of the ammonia used for cell synthesis will be returned to the liquid through lysis and autoxidation. The remaining assimilated nitrogen can be removed from the system in the net growth, or wasted biological sludge.

Under appropriate conditions, discussed in subsequent sections, ammonia nitrogen can be oxidized in a two-step process to form nitrates. This process, called nitrification, is carried out by two groups of microorganisms (nitrifiers) in the presence of oxygen. The transformation processes associated with nitrification are shown in Figure 2-2. Finally, nitrates may be transformed to nitrogen gas through a process called denitrification. This transformation is accomplished by denitrifying microorganisms in the absence of oxygen. An organic carbon source is required for denitrification to occur. The nitrogen gas formed escapes to the atmosphere. A residual of nondegradable soluble organic nitrogen of about 1 mg N/L will remain in the effluent.

2.4 Overview of Available Nitrogen Removal Options

Nitrogen entering a biological treatment system in the organic or ammonia form can be either removed or transformed to another form. Removal of nitrogen is obtained by assimilation and by conversion to nitrogen gas through nitrification and denitrification. Transformation of ammonia and organic nitrogen to the oxidized form of nitrate is accomplished through biological nitrification.

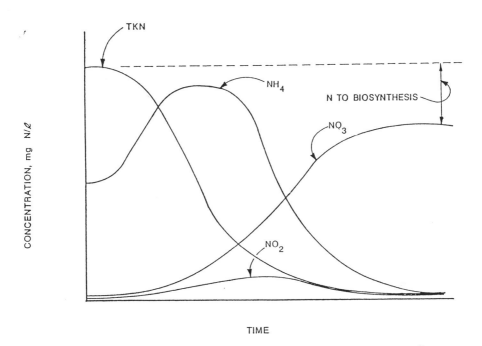

Figure 2-2. Biological nitrogen conversion processes.

5

2.4.1 Nitrogen Removal by Assimilation

Since nitrogen is an essential constituent of microbial cells, any net growth of biomass that is removed from the waste stream will cause some nitrogen removal. The amount of nitrogen that can be removed by this mechanism is limited by the amount of net growth, which in turn depends on the carbonaceous organic content of the wastewater and the system's operating conditions. Since the nitrogen content of microbial cells is approximately 12.5 percent (on a dry weight basis), the amount of nitrogen that will be removed by assimilation will be

$$dNH_3\text{-}N/dt = (0.125)\,(dX_v/dt) \tag{1}$$

where: $dNH_3\text{-}N/dt$ = rate of nitrogen removed by assimilation, lb/day
dX_v/dt = rate of active biomass or biological sludge production, lb/day

In an activated sludge system the ratio of ammonia nitrogen removed to BOD removed can be expressed by

$$\frac{dNH_3\text{-}N/dt}{dBOD/dt} = (0.125)\,\frac{dX_v/dt}{dBOD/dt} \tag{2}$$

The nitrogen content of the waste activated sludge will decrease due to endogenous metabolism. This is shown in Figure 2-3 as a function of SRT.

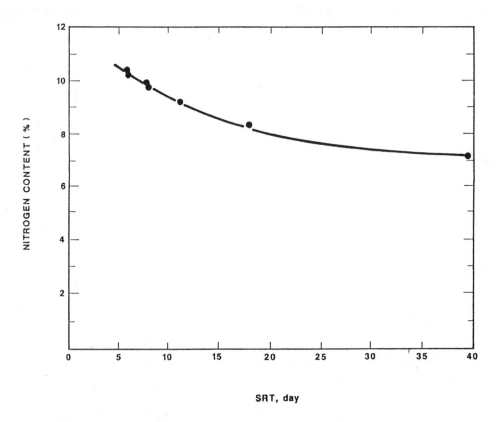

Figure 2-3. Nitrogen content of biological volatile suspended solids.

The net consumption of NH_3-N as a function of BOD removal and F/M is shown in Equation (3).

$$\frac{dNH_3\text{-}N}{dBOD} = (0.125)(a) - \frac{(0.125)(X_d)(k_b)}{F/M} \qquad (3)$$

where:
a = yield coefficient, g VSS/g BOD
X_d = degradable fraction of MLVSS
k_b = endogenous decay rate, g VSS/g VSS-day
F/M = organic loading rate, lb BOD/lb VSS-day

Since the yield coefficient, a, is typically not higher than 0.6, the theoretical maximum ratio of ammonia removed to BOD removed is 0.075. The actual ratio will be lower, depending on a system's organic loading rate (F/M). For a system operated at a F/M of 0.1 day^{-1} the ammonia to BOD removal ratio will be 0.018. Hence, nitrogen removal by assimilation is limited to approximately 2 to 5 percent of the raw wastewater BOD, depending on operating conditions. Based on primary effluent BOD and nitrogen concentrations of 120 and 30 mg/L, respectively, the percent nitrogen removal in the treatment of domestic wastewater may range from 8 to 20 percent. This removal mechanism may become quite significant in wastewaters having relatively high concentrations of BOD, such as in some industrial wastewaters or municipal wastewaters with a large industrial contributor.

Net growth should be maximized (by increasing organic loading) in order to maximize assimilative nitrogen removal. It is important to be aware that when a large portion of assimilated nitrogen returns to a waste stream from sludge handling processes, particularly heat treatment and anaerobic digestion, the overall nitrogen removal will be less.

2.4.2 Nitrification

Nitrification is the biological oxidation of ammonia to nitrate with nitrite formation as an intermediate. The microorganisms involved are the autotrophic species **Nitrosomonas** and **Nitrobacter** which carry out the reaction in two steps:

$$2\,NH_4^+ + 3\,O_2 \xrightarrow{\text{Nitrosomonas}} 2\,NO_2^- + 2\,H_2O + 4\,H^+ + \text{new cells}$$

$$2\,NO_2^- + O_2 \xrightarrow{\text{Nitrobacter}} 2\,NO_3^- + \text{new cells}$$

Since a buildup of nitrite is rarely observed (see Figure 2-2), it can be concluded that the rate of conversion to nitrate controls the rate of overall reaction.

The extent of nitrification that occurs during treatment is dependent on the extent to which nitrifying organisms are present. The cell mass comprised of nitrifying organisms is referred to here as the nitrifier's volatile suspended solids(NVSS). The cell yield for **Nitrosomonas** has been reported as 0.05 - 0.29 g NVSS/g NH_3-N and for **Nitrobacter** 0.02 - 0.08 g NVSS/g NO_2-N. A value of 0.15 g NVSS/g NH_3-N oxidized is usually used for design purposes (2). The empirical overall reaction including oxidation and synthesis is:

$$NH_4^+ + 1.83\,O_2 + 1.98\,HCO_3^- \rightarrow 0.98\,NO_3^- + 0.021\,C_5H_7NO_2 + 1.88\,H_2CO_3 + 1.04\,H_2O$$

7

Thus, the stoichiometric equation for nitrification indicates that for one gram of ammonia nitrogen removed approximately:

4.33 g of O_2 are consumed	7.14 g of alkalinity (as $CaCO_3$) are destroyed
0.15 g of new cells are formed	0.08 g of inorganic carbon are consumed

In wastewaters with low alkalinity and/or high ammonia concentrations, alkalinity may have to be added in order to maintain the pH at the optimum level for nitrification. Lime or bicarbonate can be used for this purpose.

The effect of pH on the nitrification reaction is shown in Figure 2-4(3). As shown in the figure, over the range of pH 7.0 to 8.0 there is little effect on nitrification rate. Since the pH of municipal wastewater usually falls within this range, pH should not be a factor.

However, it is especially important that there be sufficient alkalinity in the wastewater to balance the acid produced by nitrification or else alterations in pH could have an adverse effect on nitrification. As indicated above, about 7.14 mg of alkalinity (as $CaCO_3$) are consumed per mg NH_3-N oxidized. This means that municipal wastewater with 50 mg TKN/L available for oxidation should have an alkalinity of about 400 mg/L (as $CaCO_3$) to ensure a residual of 40 mg/L (as $CaCO_3$) after full nitrification. The consumption of alkalinity has a depressing effect on the pH. Nitrification reduces the HCO_3^- concentration and increases the H_2CO_3 concentration. This effect is mediated by CO_2 stripping duration aeration. If the CO_2 is not stripped from the liquid, as can occur in high purity oxygen systems, the alkalinity may have to be as much as 10 times greater than the amount of ammonia nitrified. As shown later, this problem is alleviated when denitrification is employed since one-half of the alkalinity is recovered in the denitrification process.

In all domestic and in most industrial wastewaters, the concentration of carbonaceous organics greatly exceeds that of nitrogen. The heterotrophic organisms yield also exceeds that of the autotrophs. Hence, the autotrophic population normally constitutes a small fraction of the total biomass. Neglecting the endogenous decay process, the nitrifier's (autotrophs) fraction can be estimated by

$$F_N = \frac{(a_N)(A_r)}{(a)(S_r) + (a_N)(A_r)} \qquad (4)$$

where:
F_N = nitrifier fraction
a_N = nitrifier yield coefficient, g NVSS/g NH_3-N
a = heterotrophs yield coefficient, g VSS/g BOD
A_r = ammonia nitrogen removed, mg/L
S_r = BOD removed, mg/L

In order to maintain a population of nitrifying organisms in a mixed culture of activated sludge, the system sludge age, or solids retention time(SRT), must exceed the reciprocal of the nitrifiers' net specific growth rate. This was shown by Downing et al.(4) and can be expressed by

$$SRT = \frac{1}{\mu_N - k_{Nd}} \qquad (5)$$

where:
SRT = system solids retention time or sludge age, days
μ_N = nitrifiers specific growth rate, day^{-1}
k_{Nd} = nitrifiers decay rate, g NVSS/g NVSS-days

8

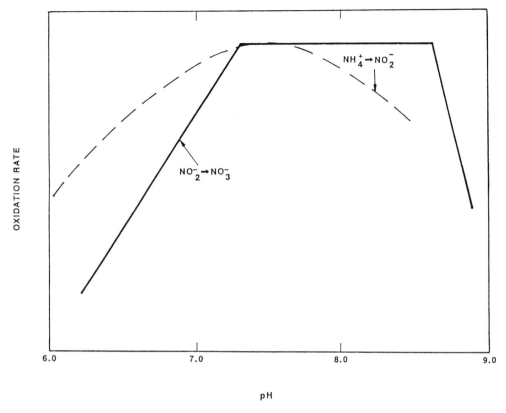

Figure 2-4. Effect of pH on ammonia oxidation.

Hence, the minimum solids retention time or sludge age required for nitrification is

$$SRT_{min} = \frac{1}{\mu_{N,\,max}} \qquad (6)$$

where: SRT_{min} = minimum solids retention time required for nitrification, days
$\mu_{N,\,max}$ = maximum specific growth rate of nitrifiers, g NVSS/g NVSS-day

The actual growth rate of nitrifiers in an activated sludge system is dependent on the concentrations of ammonia nitrogen and DO, as well as the system pH. The effects of DO and effluent ammonia are expressed by a Monod Kinetic expression:

$$\mu_N = (\mu_{N,\,max}) \left[\frac{NH_3\text{-}N}{K_N + NH_3\text{-}N} \right] \left[\frac{DO}{K_O + DO} \right] \qquad (7)$$

where K_N and K_O are the half saturation coefficients for nitrogen and oxygen, respectively. Typical values of K_N and K_O are 0.5 and 0.3 mg/L, respectively[2]. K_O has been reported to vary from 0.2 to 1.0. The influence of dissolved oxygen on nitrification rates has been somewhat controversial, partly because the bulk liquid concentration is not necessarily the same as that inside the floc where the oxygen is consumed. Also, in full-scale systems with mechanical aeration the oxygen concentration varies spatially due to oxygen being introduced to the wastewater at discrete points.

9

The effects of oxygen concentration on the specific growth rate of **Nitrosomonas** should be considered when combined carbon/nitrogen removal processes are used. In such systems the nitrifying bacteria may comprise only about 5 percent of the total biomass. Increased oxygen concentrations would increase the penetration of oxygen into the floc, thereby increasing the rate of nitrification. At a decreased SRT, the oxygen utilization rate due to carbon oxidation increases, thereby decreasing the penetration of oxygen. Conversely at a high SRT, the low oxygen utilization rate permits oxygen penetration even at low dissolved oxygen levels and consequently high nitrification rates occur. Therefore, in order to maintain maximum nitrification, the dissolved oxygen concentration would have to be increased as the SRT was decreased. This is schematically shown in Figure 2-5. For case (a) at a low F/M ratio and low dissolved oxygen concentration, the entire floc is aerobic and nitrification proceeds at a maximum rate. In case (b) the higher oxygen utilization rate resulting from a higher F/M ratio decreases the penetration of oxygen and the nitrification rate is suppressed. Increasing the dissolved oxygen concentration at a higher F/M ratio (case (c)) permits greater oxygen penetration and an increased nitrification rate.

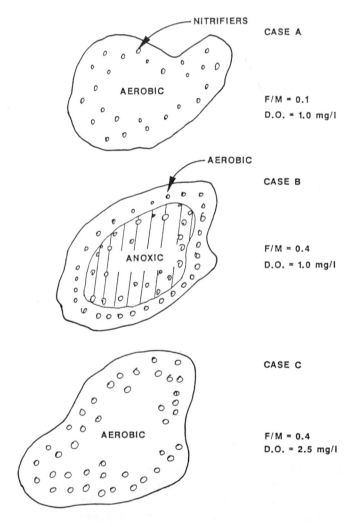

Figure 2-5. Effect of F/M on nitrification.

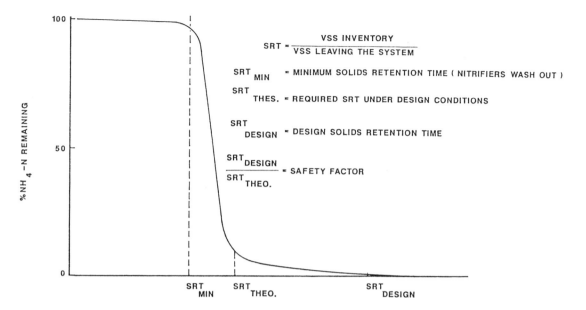

Figure 2-6. Relationship of ammonia removal and solids retention time in an activated sludge system.

The maximum growth rate of nitrifiers at 15°C is approximately 0.45 day^{-1}(2), and its temperature dependence is given by

$$\mu_{N,\ max(T)} = (0.45)\ (e^{0.098(T-15)}) \tag{8}$$

$\mu_{N,max}$ has been reported to vary from 0.46 to 2.2(5).

The theoretical sludge age required for nitrification is obtained by substituting equations (7) and (8) into equation (5). This will give the steady-state sludge age required in order to achieve nitrification at given operating NH_3-N and DO levels.

Due to diurnal variations in raw waste loads a safety factor is applied to the theoretical minimum SRT to obtain the design SRT:

$$SRT_{design} = (SRT_{theor})\ (SF) \tag{9}$$

The safety factor may equal the peak to average nitrogen load ratio, typically in the range of 1.5 to 2.5. These relationships are shown in Figure 2-6.

The SRT can be determined:

$$SRT = (X_v)(t)/dX_v \tag{10}$$

where: X_v = mixed liquor volatile suspended solids concentration, mg/L
  dX_v = volatile suspended solids (including effluent VSS) wasted per day, mg/L
  t = hydraulic detention time, day

11

Many processes today include anoxic zones for denitrification. Since the nitrifiers can only grow in the presence of oxygen, no growth can take place in the unaerated zones. Furthermore, the endogenous breakdown takes place in both the aerated and unaerated zones. Therefore, it is usual to base the criteria of minimum SRT on the characteristics of the aerated section of the treatment plant.

Equation (10) can be reexpressed:

$$SRT = \frac{X_v t}{(a)(S_r) - (k_b)(X_d)(X_v)(t)} \qquad (11)$$

k_b can be corrected for temperature:

$$k_b(T) = (k_{b(20°C)})(1.04^{(T-20)}) \qquad (12)$$

For the required SRT, $X_v t$ is computed. For a plant design, X_v is usually selected and t computed. For a plant retrofit in which t is defined by the existing wastewater flow, X_v is computed. The required oxygen is computed by:

$$O_2, mg/L = (4.33)(N_{oxidized}, mg/L)$$

The required alkalinity is computed by:

$$Alkalinity, mg/L = (7.14)(N_{oxidized}, mg/L)$$

The specific rate of nitrification can be expressed by

$$q_N = \mu_N/a_N \qquad (13)$$

where: q_N = specific rate of nitrification, g NH_3-N/g NVSS - day

The ammonia removal rate (or nitrification rate) is given by

$$R_N = (q_N)(X_N) \qquad (14)$$

where: R_N = ammonia removal rate, mg/L/day
X_N = concentration of nitrifiers, mg/L

This rate can be expressed in terms of the total biomass concentration and the nitrifier fraction:

$$R_N = (q_N)(F_N)(X_v) \qquad (15)$$

The nitrification capacity of a system equals the product of the nitrogen removal rate (R_N) and the detention time. Transient peak loading conditions allowable are the sum of a system's nitrification capacity and the allowable discharge concentration.

Example 1. Determine the design sludge age, or SRT, for nitrification under the following conditions:

Water temperature = 10°C
k_{Nd} = 0.05 day^{-1}
Safety factor = 2.0
Heterotrophic yield coefficient, a = 0.55
X_d = 0.64

Effluent ammonia = 1.5 mg/L as N
Aeration basin DO = 2.0 mg/L
BOD removed = 200 mg/L
k_b = 0.1 at 20°C
a_N = 0.15

Solution. The maximum specific growth rate for nitrifiers at 10°C is (from equation 8)

$$\mu_{N,\,max(10°)} = (0.45)\,(e^{0.098(10-15)})$$

$$= (0.45)\,(0.613)$$

$$= 0.276\ day^{-1}$$

The nitrifier specific growth rate at the specified condition is (from equation 7)

$$\mu_{N(10°)} = 0.276\,[\,\frac{1.5}{0.5+1.5}\,]\,[\,\frac{2.0}{0.3+2.0}\,]$$

$$= (0.276)\,(0.75)\,(0.87)$$

$$= 0.180\ day^{-1}$$

The theoretical sludge age required is (from equation 5)

$$SRT_{theor} = \frac{1}{0.180 - 0.05}$$

$$= 7.7\ days$$

The design sludge age is (from equation 9)

$$SRT_{design} = (2.0)\,(7.7)$$

$$= 15.4\ days$$

The MLVSS and t are determined (from equations 12 and 11):

$$k_{b(10°C)} = (0.1)\,(1.04^{(10-20)})$$

$$= 0.067$$

$$SRT = \frac{X_v t}{(0.55)(200) - (0.067)(0.64)(X_v t)}$$

For SRT = 15.4 days, $X_v t$ = 1,020. If the residence time is taken to be 12 hours (0.5 day), the required MLVSS is 2,040 mg/L.

13

Example 2. For the system of Example 1, the average influent TKN is 30 mg/L. Calculate the peak transient influent TKN concentration allowed for the effluent NH_3-N to remain below 1.5 mg/L. Neglect ammonia assimilation.

Solution. At an operating effluent NH_3-N concentration of 1.5 mg/L, the nitrifier growth rate is 0.180 day^{-1}, as shown in Example 1. Hence, the specific nitrification rate is (from equation 13)

$$q_N \quad = \quad 0.180/0.15$$

$$= \quad 1.20 \text{ g } NH_3\text{-N/g NVSS - day}$$

The nitrifier fraction is estimated by (from equation 4):

$$F_N \quad = \quad \frac{(0.15)\,(30)}{[(0.55)\,(200)] + [(0.15)\,(30)]}$$

$$= \quad 0.039$$

The nitrogen removal rate is (from equation 15)

$$R_N \quad = \quad (1.20)\,(0.039)\,(2,040)$$

$$= \quad 95.5 \text{ mg/L - day}$$

At a residence time(t) of 12 hr (0.5 day) the plant's nitrification capacity is

$$(R_N)(t) \quad = \quad (95.5)\,(0.5)$$

$$= \quad 47.7 \text{ mg N/L}$$

The allowed peak transient ammonia load, which may last several hours, is

$$= \quad 47.7 + 1.5$$

$$= \quad 49.2 \text{ mg N/L}$$

2.4.2.1 Factors Affecting Nitrification

Nitrifying organisms are subject to inhibition by various organic compounds. Hockenbury and Grady(6) have summarized inhibition data for selected organic compounds as shown in Tables 2-2 and 2-3.

If inhibitory compounds are present in wastewater, then the performance of separate stage or combined systems will probably be different. In a separate stage system, the inhibitory substance will probably be degraded in the first stage and second stage kinetics will proceed normally.

The performance of a combined system may be poorer because of reduced degradation of the inhibitory substance. If the SRT is high in a completely mixed activated sludge system (CMAS), then the inhibitory substance will most probably degrade and nitrification will proceed normally. However, in the case of a plug flow or multi-stage CMAS, the concentration of the inhibitory substance at the front end of the process could be sufficiently high enough to inhibit nitrification completely. In this case, nitrification will not proceed until the inhibitory substance is degraded and, therefore, only a portion of the SRT is available for growth of the nitrifiers. This implies that a longer SRT would be required in these cases.

Table 2-2. Effects of organic compounds on degree of inhibition of ammonia oxidation.

Compound	Degree of inhibition at the concentration (mg/L) indicated, %				Estimated concentration giving 50% inhibition, mg/L
	100	50	10	As noted	
Dodecylamine	96	95	--	66[b]	<1
Aniline[a]	86	--	--	76[c], 89[d]	<1
n-Methylaniline	90	83	71	--	<1
1-Naphthylamine	81	81	45	--	15
Ethylenediamine[a]	73	--	41	61[e]	17
Napthylethylenediamine diHCl	93	79	29	--	23
2,2'-Bipyridine	91	81	23	--	23
p-Nitroaniline	64	52	46	--	31
p-Aminopropiophenone	80	56	22	--	43
Benzidine diHCl	84	56	12	--	45
p-Phenylazoaniline	54	47	0	--	72
Hexamethylene diamine[a]	52	45	27	--	85
p-Nitrobenzaldehyde	76	32	29	--	87
Triethylamine	35	--	--	63[f]	127
Ninhydrin	30	26	31	--	>100
Benzocaine	30	27	0	--	>100
Dimethylgloxime	30	9	--	56[f]	140
Benzylamine	26	10	0	--	>100
Tannic acid	20	7	--	22[f]	>150
Monoethanolamine[a]	16	--	--	20[g]	>200

[a]Industrially significant chemicals
[b]1 mg/L
[c]2.5 mg/L
[d]5 mg/L
[e]30 mg/L
[f]150 mg/L
[g]200 mg/L

15

Table 2-3. Organic compounds that inhibit activated sludge nitrification.

Compound	Concentration[a] (mg/L)
Acetone[b]	2000.000
Allyl alcohol	19.500
Allyl chloride	180.000
Allyl isothiocyanate	1.900
Benzothiazole disulfide	38.000
Carbon disulfide[b]	35.000
Chloroform[b]	18.000
o-Cresol	12.800
Diallyl ether	100.000
Dicyandiamide	250.000
Diguanide	50.000
2,4-Dinitrophenol	460.000
Dithiooximide	1.100
Ethanol[b]	2400.000
Guanidine carbonate	16.500
Hydazine	58.000
8-Hydroxyquinoline	72.500
Mercaptobenzothiazole	3.000
Methylamine hydrochloride	1550.000
Methyl isothiocyanate	0.800
Methyl thiuronium sulfate	6.500
Phenol[b]	5.600
Potassium thiocyanate	300.000
Skatol	7.000
Sodium dimethyl dithiocarbamate	13.600
Sodium methyl dithiocarbamate	0.900
Tetramethyl thiuram disulfide	30.000
Thioacetamide	0.530
Thiosemicarbazide	0.180
Thiourea	0.076
Trimethylamine	118.000

[a]Concentration giving approximately 75% inhibition.
[b]Industrially significant chemicals.

16

In many industrial wastewaters or municipal wastewaters with a high industrial input, the rate of nitrification is sharply reduced. Figure 2-7 compares the nitrification rate for a coke plant wastewater to municipal sewage at various temperatures. In some cases where inhibition is present, the addition of powdered activated carbon (PAC) has enhanced nitrification as shown in Table 2-4. Anthonisen(8) has shown toxicity of the nitrification process can occur due to ammonia or nitrite. Since only ammonia and nitrite in the un-ionized form are toxic, these effects are a function of pH, as shown in Figure 2-8. The optimal pH for nitrification varies between 6 and 7.5, depending on the formation of free ammonia and free nitrous acid.

Table 2-4. Effect of PAC on nitrification of coke plant wastewaters(8).

PAC Feed (mg/L)	SRT (d)	TOC (mg/L)	TKN (mg/L)	NH_3-N (mg/L)	NO_2-N (mg/L)	NO_3-N (mg/L)
0	40	31	72	68	4.0	0
33	30	20	6.3	1	4.0	9.0
50	40	26	6.4	1	1.0	13.0

Influent conditions: TOC = 535 mg/L, TKN = 155 gm/L, NH_3-N = 80 mg/L

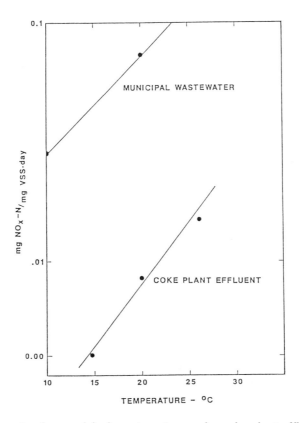

Figure 2-7. Nitrification rates for municipal wastewaters and a coke plant effluent.

17

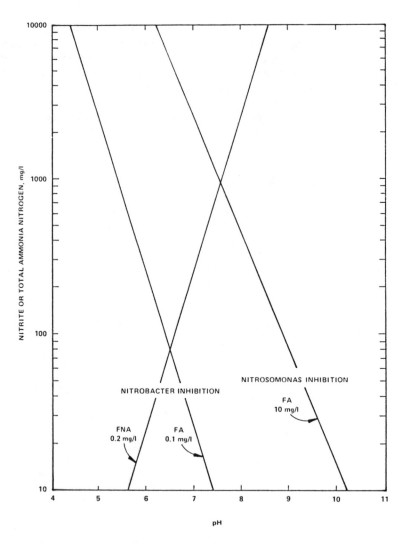

Figure 2-8. Ammonia and nitrite inhibition to nitrification (FA = Free Ammonia; FNA = Free Nitrous Acid).

2.4.3 Denitrification

Denitrification is the biological conversion of nitrate-nitrogen to more reduced forms such as N_2, N_2O and NO. The process is brought about by a variety of facultative heterotrophs which can utilize nitrate instead of oxygen as the final electron acceptor. It was shown that the breakdown of carbonaceous organics in the denitrification process is similar to that in the aerobic process, the only difference being in the final stages of the electron transfer. Thus, the term anoxic denitrification would seem more appropriate than anaerobic denitrification. This would indicate the need for strict anoxic conditions in a denitrifying system. However, it has been shown that under acidic pH conditions denitrification can take place in the presence of oxygen. Moreover, fixed film reactors, as well as suspended growth systems, may consist of aerobic biomass layers and anoxic sublayers so that aerobic processes and denitrification may occur simultaneously.

18

The stoichiometric reaction describing denitrification depends on the carbonaceous matter involved. For methanol, which is the most extensively used and studied external carbon source, the reaction is:

$$6 \, NO_3^- + 5 \, CH_3OH \rightarrow 3 \, N_2 + 5 \, CO_2 + 7 \, H_2O + 6 \, OH^-$$

Including cell synthesis the empirical reaction is:

$$NO_3^- + 1.08 \, CH_3OH + 0.24 \, H_2CO_3 \rightarrow 0.06 \, C_5H_7NO_2 + 0.47 \, N_2 + 1.68 \, H_2O + HCO_3^-$$

This reaction expression indicates that for one gram of nitrate-nitrogen that is denitrified:

   2.47 g of methanol (or approximately 3.7 g of COD) are consumed
   0.45 g of new cells are produced
   3.57 g of alkalinity are formed

Nitrate will also replace oxygen in the endogenous respiration reaction. The proposed equation is:

$$C_5H_7NO_2 + 4.6 \, NO_3^- \rightarrow 5 \, CO_2 + 2.8 \, N_2 + 4.6 \, OH^- + 1.2 \, H_2O$$

The rate of denitrification depends primarily on the nature and concentration of the carbonaceous matter undergoing degradation. Most investigators agree that denitrification is a zero order reaction with respect to nitrate down to very low nitrate concentration levels[9]. Hence the nitrate removal in an anoxic basin when carbon is not limiting can be expressed by

$$(NO_3^-)_o - (NO_3^-)_e = (R_{DN}) \, (X_v)(t) \tag{16}$$

where: $(NO_3^-)_o$, $(NO_3^-)_e$ = influent and effluent nitrate nitrogen, respectively, mg/L
R_{DN} = zero order rate of denitrification, g NO_3-N/g VSS-day

Values of R_{DN} for various carbon sources are given in Table 2-5.

Table 2-5. Denitrification rates with various carbon sources.

Carbon Source	Denitrification Rate (g NO_3-N/g VSS-day)	Temperature (°C)	References
Methanol	0.21 to 0.32	25	9
Methanol	0.12 to 0.90	20	2
Sewage	0.03 to 0.11	15-27	2
Sewage	0.072 to 0.72[a]	--	11
Endogenous Metabolism	0.017 to 0.048	12-20	2

[a]the high value is for the readily biodegradable organics of raw sewage.

The rate of denitrification is dependent on temperature and DO concentration:

$$R_{DN(T)} = R_{DN(T)} K^{(T-20)} (1 - DO) \qquad (17)$$

Values of K range from 1.03 to 1.1. A value of 1.09 is commonly used. The denitrification rate will depend on both the concentration and the biodegradability of the carbon source.

Analogous to oxygen utilization in aerobic systems, the denitrification rate can be expressed by:

$$(NO_3\text{-}N)_r = (A'_N) (S_r) + (b'_N) (X_{vt}t) \qquad (18)$$

where: A'_N = nitrate utilization in anoxic degradation, g NO_3-N/g BOD
b'_N = nitrate utilization in endogenous respiration under anoxic conditions, g NO_3-N/g VSS-day

Equation (18) can be rearranged:

$$\frac{(NO_3\text{-}N_r)}{X_{vt}t} = A'_N \frac{S_r}{X_{vt}t} + b'_N$$

This relationship is shown in Figure 2-9.

Treating municipal wastewater, Barnard (10) found three distinct denitrification rates, as shown in Figure 2-10. The first rate of 50 mg/L-h lasted from 5 to 15 minutes and was attributed to by-products from anaerobic fermentation. The second rate of 16 mg/L-h was attributed to normal assimilation of the particulate and more complex compounds and lasted until all external food sources were exhausted. The third rate of 5.4 mg/L-h was attributed to endogenous respiration.

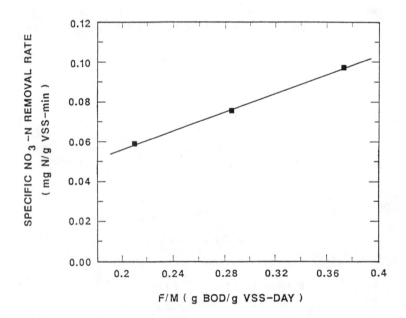

Figure 2-9. Denitrification rate as a function of F/M.

20

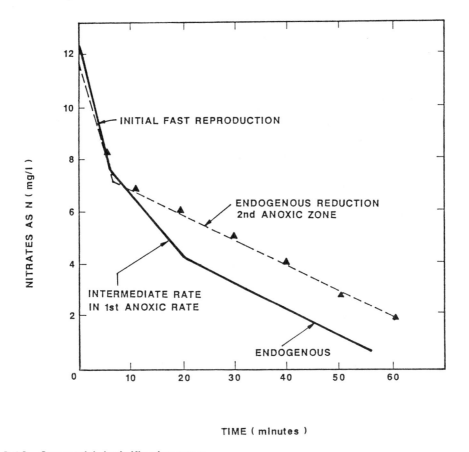

TIME (minutes)

Figure 2-10. Sequential denitrification rates.

Ekama and Marais (11) produced the following models for these three rates:

$$R_{DN(1)} = 0.72\ \theta_1^{(T-20)}\ X_a$$
$$R_{DN(2)} = 0.101\ \theta_2^{(T-20)}\ X_a$$
$$R_{DN(3)} = 0.072\ \theta_3^{(T-20)}\ X_a$$

in which $\theta_1 = 1.20$, $\theta_2 = 1.03$, and $\theta_3 = 1.03$. X_a is defined as the active mass, i.e. $X_d X_v$.

The denitrification rate under aerobic conditions will depend on the anoxic fraction of the biological floc and the availability of carbon substrate. The DO term in equation (17) indicates that the denitrification rate decreases linearly to zero when the dissolved oxygen concentration reaches 1.0 mg/L. Further research is necessary tho establish the actual denitrification rate versus dissolved oxygen levels. Denitrification rates of 0.006 mg NO_3-N/mg VSS/day have been reported under aerobic conditions(12). For practical purposes, denitrification can be ignored when dissolved oxygen concentrations are greater than 1.0 mg/L.

Recent experience in South Africa has shown significant denitrification occurring in an anaerobic plug flow process in which surface aerators are used and the surface dissolved oxygen maintained at 1.0-2.0 mg/L. The hypothesis is that in the lower levels of the tank denitrification occurs at reduced oxygen levels.

Example 3. Calculate the residence time required for denitrification in an anoxic basin in an activated sludge plant under the following conditions:

Influent nitrate = 25 mg N/L $R_{DN(20°)}$ = 0.10 day^{-1}
Effluent nitrate = 5 mg N/L   K = 1.09
Temperature = 10°C   DO = 0.1 mg/L
MLVSS = 2,000 mg/L

Solution. The denitrification rate at 10°C is (from equation 17):

$$R_{DN(10°)} = (0.10) (1.09^{(10-20)}) (1 - 0.1)$$

$$= (0.10) (0.42) (0.9)$$

$$= 0.038 \text{ g } NO_3\text{-N/g VSS-day}$$

The required residence time is (from equation 16):

$$t = \frac{25 - 5}{(2,000) (0.038)}$$

$$= 0.263 \text{ day}$$

$$= 6.3 \text{ hr}$$

2.4.4 Combined Nitrification/Denitrification

Removal of nitrogen can be accomplished through denitrification of nitrified wastewater. The process can take one of three basic forms: (a) two-sludge or separate stage system (Figure 2-11a) (b) single-sludge system with mixed liquor recycle (Figure 2-11b), and (c) an oxidation ditch or channel in which nitrification and denitrification occur sequentially (Figure 2-11c). In addition, several modifications of these basic processes have been proposed. In the two-sludge system carbonaceous organic removal and nitrification take place in the first aerobic activated sludge unit. The clarified effluent of this stage is passed to the second stage where anoxic conditions prevail and denitrification occurs. Since the organic carbon of the raw wastewater has been largely removed in the first stage, an external carbon source (e.g., methanol) is required to serve as electron acceptor in the denitrification basin.

In the single-sludge recycle system, the mixed liquor contains a mixture of heterotrophic and autotrophic microorganisms. The heterotrophs grow and oxidize carbonaceous organics in both the aerobic and anoxic basins. They utilize molecular oxygen as the electron acceptor in the former basin and nitrate in the latter. The autotrophs grow in the aerobic basin only, using molecular oxygen and inorganic carbon while oxidizing ammonia. The influent ammonium passes through the anoxic basin to the aerobic basin where it is converted to nitrate. The effluent from the aerobic basin is recycled to the anoxic basin where the nitrate is reduced. A key feature of this single-sludge system is the high rate of mixed liquor recycle from the aerobic to the anoxic basin (200 to 500 percent).

In the oxidation ditch, both nitrification and denitrification are occurring in the same basin through alternating aerobic and anoxic zones.

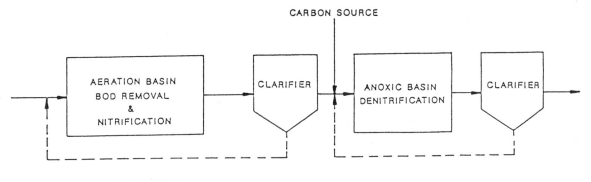

(a) TWO-SLUDGE SYSTEM

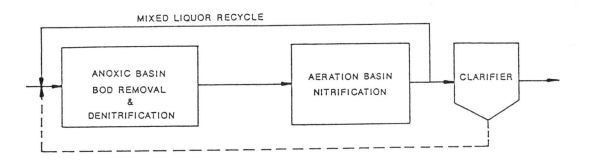

(b) SINGLE-SLUDGE (RECYCLE) SYSTEM

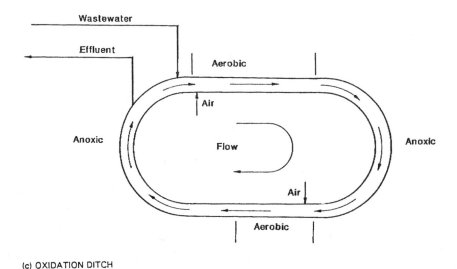

(c) OXIDATION DITCH

Figure 2-11. Alternative systems for biological nitrogen removal.

23

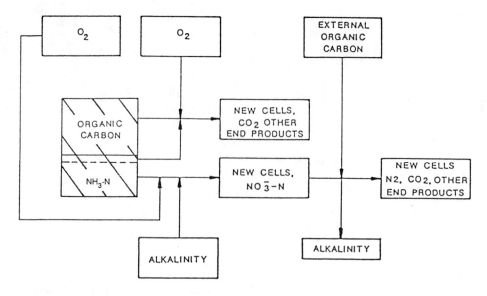

(a) TWO-SLUDGE SYSTEM

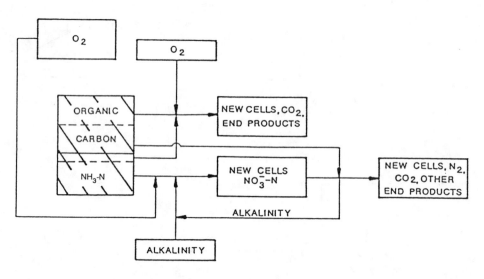

(b) SINGLE-SLUDGE (RECYCLE) SYSTEM

Figure 2-12. Schematic materials flow diagram for nitrogen removal systems.

The single-sludge recycle system offers several economical advantages over the two-sludge system in that it uses only one clarification step, no external carbon source, has lower neutralization chemical requirements and has lower oxygen requirements. Schematic materials flow diagrams shown in Figure 2-12 illustrate the differences between these two systems that contribute to these chemical and oxygen savings. In both schemes, influent ammonia is nitrified utilizing the same amount of oxygen. A small portion of the ammonia is assimilated into cellular material. As illustrated in Figure 2-12(a), the influent organic carbon is completely oxidized aerobically utilizing its equivalent amount of oxygen.

24

As illustrated in Figure 2-12(b), only a fraction of this organic carbon is oxidized aerobically in a single-sludge system, the remaining being utilized in the anoxic zone. This eliminates the need for an external carbon source, as shown in Figure 2-12(b). The potential savings in neutralizing chemicals associated with the single-sludge system is also illustrated in Figure 2-12(b), where part of the alkalinity consumed in the aerobic zone is recovered in the anoxic zone. These savings are partially offset by the need for pumping equipment and energy for recycling high volumes of mixed liquor.

Detailed design procedures for the single-sludge recycle system are presented elsewhere (13,14,15). One simplified method is presented subsequently. Other methods are available.

Assuming complete denitrification of recycled NO_3-N in the anoxic stage and neglecting nitrogen assimilation, the required recycle ratio (mixed liquor + return sludge) is given by

$$R = \left[\frac{(NH_3\text{-}N)_o - (NH_3\text{-}N)_e}{(NO_3\text{-}N)_e} \right] - 1 \qquad (19)$$

where: R = overall recycle (mixed liquor + return sludge) ratio

   $(NH_3\text{-}N)_o$, $(NH_3\text{-}N)_e$ = influent and effluent ammonia-nitrogen, respectively, mg/L.

   $(NO_3\text{-}N)_e$ = effluent nitrate-nitrogen, mg/L

The relationship defined by equation (19) is shown graphically in Figure 2-13.

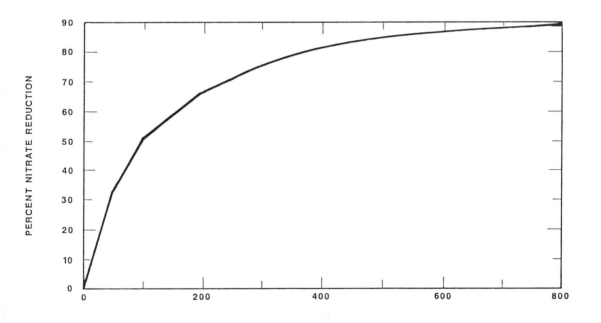

Figure 2-13. Effect of mixed liquor recycle on theoretical nitrate capture.

25

Since the nitrifiers can only grow in the aerobic zone, the minimum SRT required for nitrification can be expressed by:

$$SRT' = \frac{SRT}{V_{aerobic}} \qquad (20)$$

where: SRT' = solids retention time required for nitrification in a single-sludge recycle system, day
SRT = solids retention time required for nitrification in a conventional system, day (obtained by using equations 5, 7, 8 and 9)
$V_{aerobic}$ = aerobic volume fraction

The overall system residence time can be calculated from

$$t = \frac{(a)(S_r)(SRT')}{(X_v)\,[1 + (k_b)(X_d)(SRT')]} \qquad (21)$$

where: t = overall residence time, day
S_r = BOD removed in the system, mg/L (approximately equal to influent BOD)
X_d = degradable fraction of MLVSS under aeration

The degradable fraction of the MLVSS has been expressed by(16)

$$X_d = \frac{0.8}{1 + [(0.2)\,(k_b)\,(SRT')]} \qquad (22)$$

The anoxic residence time is calculated by

$$t_{DN} = (1 - V_{aerobic})(t) \qquad (23)$$

The required anoxic residence time for denitrification, t'_{DN}, is calculated by

$$t'_{DN} = \frac{N_{Denit}}{(R_{DN})\,(X_v)} \qquad (24)$$

Where N_{Denit} is the amount of nitrate to be denitrified, mg/L.

If $t_{DN} = t'_{DN}$ the calculation is completed. If $t_{DN} \neq t'_{DN}$, a different $V_{aerobic}$ is assumed and the calculation is repeated.

Example 4. Calculate the required aerobic and anoxic residence times and the recycle ratio for the following conditions (neglect ammonia removed by assimilation):

BOD removed = 200 mg/L   Influent TKN = 30 mg/L
Effluent ammonia = 1.5 mg/L as N   Effluent nitrate = 5 mg/L as N
Temperature = 10°C   a = 0.55 g VSS/g BOD
$k_{b(10°)}$ = 0.04 g VSS/g VSS-day $R_{DN(10°)}$ = 0.042 g NO$_3$-N/g VSS-day
DO in aeration basin = 2.0 mg/L   MLVSS = 2,500 mg/L

Solution. The required total recycle rate is (from equation 19)

$$R = \frac{30 - 1.5}{5} - 1$$

$$= 4.7$$

The required design nitrification SRT is obtained from Example 1 as 15.4 days. The following is the final step in a trial and error solution.

In the first step toward a solution, assume an aerobic volume fraction. In this instance the aerobic sludge volume is assumed to be 0.65. Thus, the overall sludge age is (from equation 20)

$$SRT' = \frac{15.4}{0.65}$$

$$= 23.7 \text{ days}$$

The degradable fraction of the MLVSS is (from equation 22)

$$X_d = \frac{0.8}{1 + [(0.2)(0.04)(23.7)]}$$

$$= 0.67$$

The overall residence time is (from equation 21)

$$t = \frac{(0.55)(200)(23.7)}{2,500(1 + [(0.04)(0.67)(23.7)])}$$

$$= 0.636 \text{ day}$$

The anoxic residence time is (from equation 23)

$$t_{DN} = (1 - 0.65)(0.636) = 0.223 \text{ day}$$

The required residence time for denitrification is (from equation 24)

$$t'_{DN} = \frac{(30 - 1.5 - 5)}{(0.042)(2,500)}$$

$$= 0.224 \text{ day}$$

Since t_{DN} and t'_{DN} are nearly equal the calculation is terminated. If they were not nearly equal, a different aerobic volume fraction would be selected and the calculations repeated.

27

2.4.5 Summary of Design Procedure for Biological Nitrification-Denitrification

Nitrification.

Data required: Influent and effluent BOD
Influent and effluent TKN
Half saturation coefficient, K_N and K_O for nitrification
Operating dissolved oxygen
Nitrifier decay rate, $K_{ND} = 0.05$ day^{-1} at 20°C
Yield coefficient, a, for BOD
Yield coefficient for nitrifiers, $a_n = 0.15$

1. Compute the maximum growth rate at the lowest operating temperature from equation (8).

    Compute the actual growth rate from equation (7).

    Compute the critical SRT from equation (5).

    Compute the design SRT by multiplying the critical SRT by a safety factor of 1.5 to 2.5.

2. Compute the product $X_v t$ from equation (11). The degradable fraction is computed from equation (22). For a new plant design, X_v is usually selected as 2,000-3,000 mg/L, and t determined. For a retrofit where t is defined, X_v is calculated.

3. Compute the nitrogen to be oxidized.

    $$N_{oxidized} = TKN_{removed} - N_{synthesized}$$

    The nitrogen synthesized is computed from equation (3).

    The nitrification rate is computed from equation (15) in which the fraction of nitrifiers is determined from equation (4).

If the permit contains a daily maximum limit as well, $X_v t$ is computed from the relationship

$$X_v t = N_{oxidized}/q_N F_N$$

in which q_N is adjusted to the higher effluent level and the $N_{oxidized}$ represents the peak transient ammonia load. The larger value of $X_v t$ as calculated in Step 2 and Step 3 is that used for design.

4. The oxygen requirement for nitrification is

    $$O_2 = (4.33) (N_{oxidized})$$

5. The alkalinity requirement is

    $$Alkalinity = (7.14) (N_{oxidized})$$

<u>Denitrification.</u>

Depending on the nature of the wastewater, R_{DN} can be determined from Figure 2-10 or computed from the Ekama-Marais relationships. This rate is corrected for temperature and dissolved oxygen from equation (17). The denitrification rate is computed from equation (16). It should be noted that up to one-half of the alkalinity is recovered and that BOD is assimilated by the denitrification reaction.

2.4.6 Denitrification in Fixed Film Reactors

Denitrification in fixed film reactors can be accomplished in a variety of column configurations using various media to support the growth of denitrifiers. In all cases, oxygen must be excluded from the column and an adequate carbon source be present.

Submerged packed bed reactors use granular media (e.g., gravel) or plastic media similar to that used in trickling filters. Fluidized bed reactors typically use sand as support media. Gas filled columns use plastic media and nitrogen gas to fill the void space.

Denitrification rates in fixed film reactors depend on the concentration of biomass which is related to the specific surface area of the support media. It also depends on the nature of the carbon source, the temperature, and other environmental factors. For submerged packed beds with plastic media the reported rates range from 4 to 26 lb N removed/1,000 ft^3 - day for temperatures in the 5 to 20°C range. Fluidized beds using fine media have denitrification rates up to 1,200 lb N/1,000 ft^3-day(2).

2.4.7 Nitrification in Fixed Film Reactors

Fixed film systems such as trickling filters and rotating biological contactors (RBC) can be used to nitrify secondary effluents. The biomass which accumulates on the media surface consists of both heterotrophic and autotrophic microorganisms. The proportion of nitrifiers in this biomass reflects the relative removals of carbonaceous organics and ammonia. In a trickling filter this ratio may vary along the filter depth as the carbonaceous organics are gradually depleted.

<u>Trickling filters.</u> Data on nitrification rates and efficiencies in trickling filters are scarce and somewhat confusing. The specific growth rates of the organisms in a fixed film reactor are a function of the concentration of substrate in the liquid passing over the film. As a result, the rates decrease with distance from the top of the bed. As the organisms grow, the film thickness will reach a maximum value determined by fluid shear at which point growth must equal loss. If a wastewater contains both biodegradable organic matter and NH_3-N, growth of both the heterotrophic and autotrophic bacteria will approach the maximum rate at the top of the filter. However, because of the higher growth rate of the heterotrophs, most of the film will contain these organisms and little or no nitrification will occur. As the liquid passes through the bed the concentration of BOD decreases and a point is reached where the growth rate of the nitrifiers is sufficient with respect to the heterotrophs that they can effectively compete in the film. As the organic matter further declines the autotrophs will make up a larger proportion of the film, causing the rate of nitrification per unit area to increase. Therefore, when a wastewater contains both organic matter and NH_3-N, only a fraction of the bed height will be available for nitrification and the magnitude of that fraction will depend both on the absolute and relative concentrations of the two types of substrates. If the wastewater contains little organic matter, as in a tertiary filter application, nitrification will occur throughout the entire bed.

ORGANIC LOADING (lb TBOD/1000 sq ft/day)

Figure 2-14. Ammonia removal at 21°C in a trickling filter(17).

Recirculation may enhance nitrification by reducing the heterotroph growth rate due to dilution, thus allowing the autotrophs to more effectively compete for space in the biofilm. Oxygen diffusion is another factor which may limit the rate of nitrification in trickling filters. Recirculation which increases the dissolved oxygen concentration may cause a reduction in that limitation by increasing the DO to NH_3-N ratio.

Nitrification with respect to organic loading is shown in Figure 2-14 (17). Recent attempts have been made to evaluate trickling filter performance in a systematic and consistent way to allow use of such data for design.

From empirical data compiled by Gullicks and Cleasby(18), nitrogen removal curves have been plotted as a function of influent ammonia concentration and hydraulic loading. These curves are shown in Figure 2-15 and Figure 2-16 for temperatures of >14°C and 10°C to 14°C, respectively. By a simple trial and error procedure one can use these curves for system design.

Boller and Gujer(19), based on pilot plant studies of tertiary trickling filters, recommend a media surface loading rate of 0.4 g NH_3-N/m^2 - day for complete nitrification (effluent NH_3-N < 2.0 mg/L) at a water temperature of 10°C. Data compiled by Barnes and Bliss(5) recommend a loading range of 0.5 to 1.0 g NH_3-N/m^2 - day for plastic media filters at temperatures ranging from 10°C to 20°C.

The effect of nitrogen loading on nitrification efficiency is shown in Figure 2-17. This figure was made using pilot plant data reported by Jiumm, Yeun and Molof(20).

30

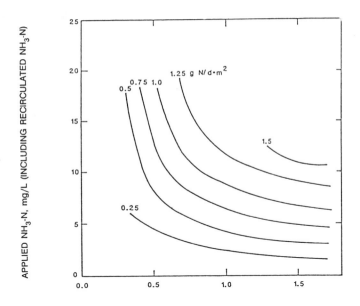

Figure 2-15. Predicted NH₃-N removals in trickling filters at T > 14°C(18).

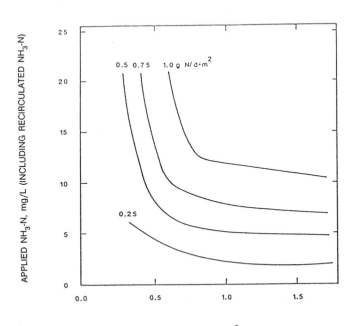

Figure 2-16. Predicted NH₃-N removals in trickling filters at T = 10 to 14°C(18).

31

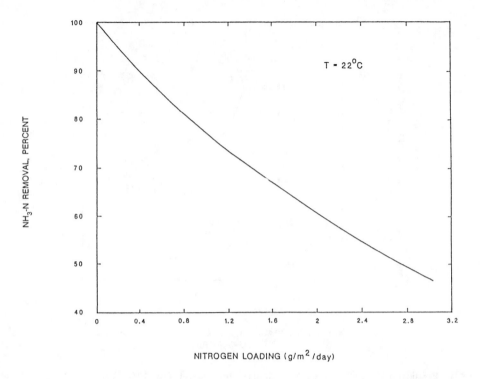

NITROGEN LOADING (g/m^2/day)

Figure 2-17. Effect of trickling filter loading on nitrogen removal in tertiary filtration on plastic media(20).

Example 5. Estimate the required volume of a plastic media trickling filter for the removal of ammonia from a secondary effluent under the following conditions:

Flow = 13.2 MGD (50,000 m^3/day)
Influent NH$_3$-N = 20 mg/L
Effluent NH$_3$-N = 2 mg/L
T = 10°C
Media specific area = 100 m^2/m^3

Solution

a) Based on Boller and Gujer recommendation:

Recommended loading = 0.4 g NH$_3$-N/m^2 - day

Ammonia to be nitrified = (50,000) (20-2) = 900,000 g/day

Required media surface area = $\dfrac{900,000}{0.4}$ = 2,250,000 m^2

Filter media volume = $\dfrac{2,250,000}{100}$ = 22,500 m^3 (794,490 ft^3)

32

(b)   Based on Figure 2-17 for 90 percent removal, the loading rate is 0.42 g/m^2 - day at 22°C. Hence for 10°C the loading rate will be adjusted using a typical coefficient for trickling filters of 1.03:

$$(0.42) \ (1.03)^{(10-22)} \ = \ 0.30 \ g/m^2 \text{ - day}$$

The loading of ammonia is 1,000,000 g/day, hence the required surface area is

$$\frac{1,000,000}{0.30} \ = \ 3,300,000 \ m^2$$

The filter media volume

$$\frac{3,300,000}{100} \ = \ 33,000 \ m^3 \ (1,165,000 \ ft^3)$$

(c)   Based on Figure 2-16 and assuming a filter depth of 7.0 m (23 ft) for removal of 0.4 g/m^2 - day and influent ammonia of 20 mg/L, the hydraulic loading is 0.17 L/sec - m^2

The required surface area $= \dfrac{1,000,000}{0.4}$ $= 2,500,000 \ m^2$

Media volume $= \dfrac{2,500,000}{100}$ $= 25,000 \ m^3 \ (883,000 \ ft^2)$

Cross-sectional area $= \dfrac{25,000}{7}$ $= 3,570 \ m^2$

Hydraulic loading $= \dfrac{(50,000) \ (10^3)}{(24) \ (3,600) \ (3,570)}$ $= 0.162 \ L/sec\text{-}m^2$

Since this is close to 0.17 L/sec - m^2, the calculation is terminated.

Rotating Biological Contactors (RBC). Biological activity in an RBC is similar to a trickling filter, in that there is sequential oxidation of the organic matter and NH_3-N in the wastewater and the degree of nitrification decreases as the BOD of the wastewater increases. Experience gained with full-scale RBC plants has shown that nitrifying bacteria cannot compete effectively for space in the biofilm until the concentration of soluble organic matter is below 15 mg BOD_5/L. However, it has been shown in many cases that maximum nitrification rates are not achieved until the soluble BOD_5 concentration is less than 5 mg/L. The accepted design procedure is to compute the surface area required to reduce the carbonaceous soluble BOD_5 to 15 mg/L and then to compute the additional surface area required for nitrification. Full-scale data are shown in Figure 2-18. The required surface area is computed in a manner similar to the removal of carbonaceous BOD. Since reactors in series are normally used, nitrification will only occur in the latter stages.

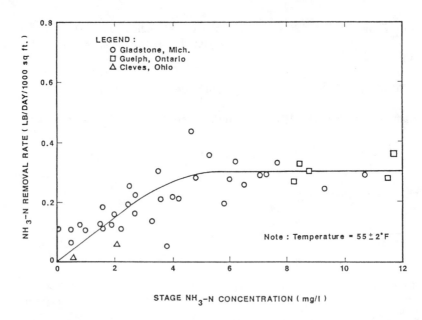

Figure 2-18. Full-scale RBC nitrification rates at design wastewater temperature (55°F) (20).

2.5 Process Selection

Process selection is affected primarily by effluent quality requirements and operational considerations. When ammonia removal is the only requirement, a nitrification system can be selected. However, such systems may experience sludge settling problems due to denitrification in the secondary clarifiers. This can be a severe problem if a raw wastewater's nitrogen levels are high or temperatures are warm. For this reason it may be desirable to include denitrification under conditions where nitrification is required. Alternatively, a fixed film nitrification reactor can be used following secondary treatment. This can be an economically feasible solution if the ammonia concentration is low and effluent nitrogen limits are not too stringent. In many cases a tertiary nitrification trickling filter does not have to be followed by a clarifier and therefore denitrification in a secondary clarifier with its associated problems would not occur.

Combined nitrification and denitrification can be achieved in a two-sludge or single-sludge (recycle) system. If low effluent nitrate levels are required, the single-sludge system can be followed by a second anoxic unit for denitrification, e.g., the Bardenpho process. The oxidation ditch configuration is particularly suited for single-sludge nitrogen removal since a high degree of recycle is obtained due to mixed liquor flow in the ditch. Zones of aerobic and anoxic conditions are developed in the ditch based on the location of aerators and feed introduction points.

2.6 Physical/Chemical Processes for Nitrogen Removal

Several physical-chemical processes have been used in the past for nitrogen removal. Although under most circumstances biological treatment is the most attractive nitrogen control technology, physical and chemical processes may be technically and economically feasible in certain situations. The major processes that fall under this category are breakpoint chlorination, selective ion exchange, and air stripping. These are discussed subsequently.

34

2.6.1 Breakpoint Chlorination

Breakpoint chlorination is accomplished by the addition of chlorine to the waste stream in an amount sufficient to oxidize ammonia-nitrogen to nitrogen gas. After sufficient chlorine is added to oxidize the organic matter and other readily oxidizable substances present, a step-wise reaction of chlorine with ammonium takes place. The reactions between the ammonium ion and chlorine leading to formation of nitrogen gas may be expressed by the following two reactions:

$$NH_4^+ + HOCl \rightarrow NH_2Cl + H_2O + H^+$$

$$NH_2Cl + 0.5\ HOCl \rightarrow 0.5\ N_2 + 0.5\ H_2O + 1.5\ H^+\ 1.5\ Cl^-$$

The overall reaction may be expressed as follows:

$$NH_4^+ + 1.5\ HOCl \rightarrow 0.5\ N_2 + 1.5\ H_2O + 2.5\ H^+ + 1.5\ Cl^-$$

Stoichiometrically, the breakpoint reaction requires a weight ratio of chlorine (expressed as Cl_2) to ammonia nitrogen at the breakpoint of 7.6:1. This is equivalent to a molar ratio of 1.5:1. In practice, the actual weight ratio of chlorine to ammonia nitrogen at breakpoint has ranged from about 8:1 to 10:1. Experience with municipal wastewater indicates that 95 to 99 percent of the influent ammonia is converted to nitrogen gas, with the remainder being nitrate (NO_3^-) and nitrogen trichloride (NCl_3).

The breakpoint chlorination curve (Figure 2-19) illustrates the reactions that occur under varying chlorine/ammonia ratios.

o   In Zone 1, the major reaction is the formation of monochloramine. The peak of the breakthrough curve theoretically occurs at a molar ratio of 1:1, or a weight ratio of 5:1 between added chlorine (expressed as Cl_2) and initial ammonia-nitrogen.

o   In Zone 2, oxidation results in the formation of dichloramine and oxidation of ammonia which reduces both residual chlorine and total ammonia concentrations. At the breakpoint the theoretical ratio of chlorine to ammonia-nitrogen is 7.6:1 (molar ratio of 1.5:1) and the ammonia concentration is at a minimum.

o   After the breakpoint (Zone 3) free chlorine residual, as well as small quantities of dichloramine, nitrogen trichloride and nitrate, increase.

The optimum pH to minimize formation of nitrogen trichloride and nitrate is near pH 7.0. Stoichiometrically, 14.3 mg/L of alkalinity is required for each 1.0 mg/L NH_3-N expected to be consumed. There will be an increase in total dissolved solids (TDS) in the effluent due to the chloride ions and neutralization. The chlorination will result in 6.2 mg TDS/L/mg NH_3-N/L oxidized. Neutralization with lime (CaO) results in a total of 12.2 mg TDS/L/mg NH_3-N oxidized/L. For example, if a wastewater contained 20 mg/L ammonia nitrogen, chlorine in a gaseous form would result in a 124 mg/L increase in TDS. Neutralization with lime (CaO) would result in a total increase of 244 mg TDS/L.

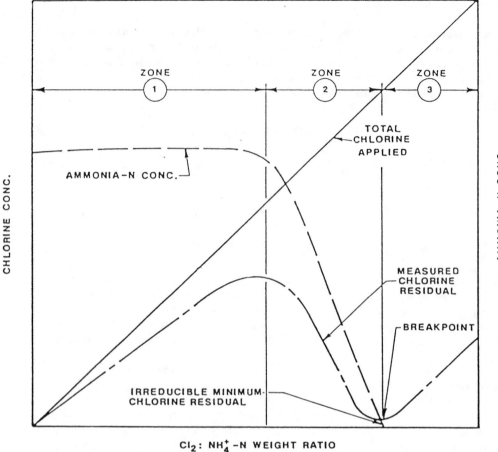

Figure 2-19. Theoretical breakpoint chlorination curve(2).

At a pH of 6 to 7 the breakpoint reaction is completed in less than 15 sec. Dechlorination will usually be required. The most common techniques involve the use of sulfur dioxide or activated carbon. In practice 0.9 to 1.0 parts of SO_2 are required to dechlorinate 1.0 part of Cl_2:

$$SO_2 + HOCl + H_2O \rightarrow Cl^- + SO_4^= + 3\,H^+$$

The resulting acidity is seldom a problem in practice due to the low concentrations involved. About 2 mg $CaCO_3$/L alkalinity are consumed for each mg SO_2/L applied.

When using activated carbon the reaction is:

$$C + 2\,HOCl \rightarrow CO_2 + 2\,H^+ + 2\,Cl^-$$

Activated carbon is expensive and should be considered only in those cases where functions other than chlorine residual control are important.

36

2.6.2 Air Stripping of Ammonia

In a wastewater stream ammonium ions exist in equilibrium with ammonia:

$$NH_3 + H_2O \rightleftharpoons NH_4^+ + OH^-$$

At pH 7 only ammonium ions (NH_4^+) exist in solution while at pH 12 the solution contains NH_3 as a dissolved gas. The relative percentages of ammonium ions and ammonia at different pH levels and temperatures are shown in Figure 2-20.

Air stripping of ammonia consists of raising the pH of the wastewater to pH 10.5 to 11.5 and providing sufficient air-water contact to strip the ammonia gas from solution. Conventional cooling towers have generally been employed for the stripping process. pH adjustment of the wastewater may employ caustic or lime. If lime is used with municipal wastewater the values shown in Figure 2-21 should generally approximate the lime requirements.

Under turbulent conditions in a stripping tower the theoretical air requirements per unit of water can be calculated from Henry's Law assuming that the air leaving the tower is in equilibrium with the influent water and that the air entering the bottom of the tower is free of ammonia. The equilibrium relationship is shown in Figure 2-22. For example, at 20°C the theoretical gas/liquid ratio can be calculated from Figure 2-22 as 1.83 moles air/mole H_2O or 305 scf/gal.

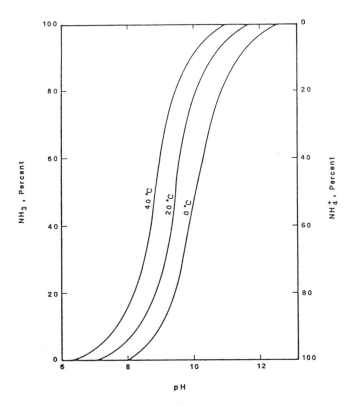

Figure 2-20. Distribution of ammonia and ammonium ion with pH and temperature.

37

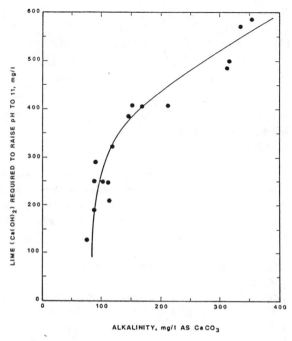

Figure 2-21. Lime required to raise the pH to 11 as a function of raw wastewater alkalinity(22).

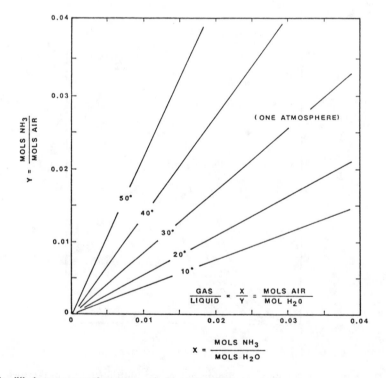

Figure 2-22. Equilibrium curves for ammonia in water(22).

38

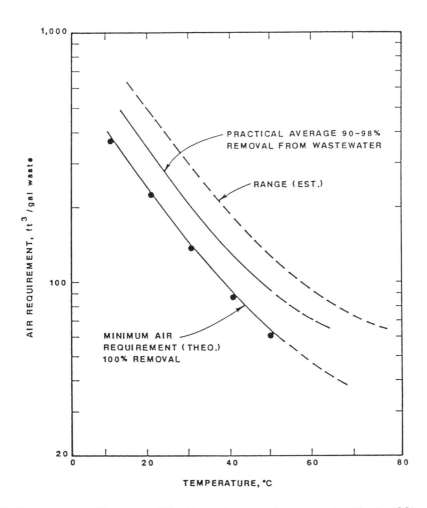

Figure 2-23. Temperature effects on air/liquid requirements for ammonia stripping(22).

In practice, Tchobanoglous(22) estimated the air requirements as shown in Figure 2-23. Ammonia removal from domestic wastewater in excess of 90 percent was found to occur at 480 ft^3/gal above pH 9.0. Hydraulic loading to stripping towers over the range 1 to 3 gpm/ft^3 (0.04 to 0.12 m^3/min-m^2) is recommended. Tower depth and packing configuration will also affect performance.

Problems associated with ammonia stripping are reduced efficiency and ice formation in colder climates, deposition of calcium carbonate on the media when lime is used for pH adjustment, possible air pollution problems and deterioration of wood packing. A process alternative has been developed in which the exhaust air from the stripper is passed through H_2SO_4 and recycled. In this way air pollution problems are eliminated, ammonium sulfate is recovered and air temperatures are maintained high.

Considerable work has been done employing ammonia stripping ponds in Israel(23). In unaerated ponds, ammonia was reduced 50 percent at a pH of 10.5 over a period of 130 hr. When aerated the retention period necessary to reduce the ammonia 50 percent was reduced to 9 to 16 hr. Post ponding resulted in a pH reduction through natural recarbonation.

2.6.3 Selective Ion Exchange

Selective ion exchange for removal of ammonia can be accomplished by passing the wastewater through a bed of ion-exchanger which exhibits a high selectivity for the ammonium ion over other cations that are normally present in wastewater. The natural zeolite clinophlolite has been found suitable for this application. It has a high selectivity for the ammonium ion with a total exchange capacity of approximately 2 meq/g. Other synthetic zeolites with considerably higher capacities are available but have not been applied in wastewater treatment plants. Regeneration of the zeolite is required when all exchange sites are utilized and ammonium breakthrough occurs.

Filtration prior to ion exchange is usually required to prevent fouling of the zeolite. Ammonium removals of 90 to 97 percent can be expected. Nitrite, nitrate, and organic nitrogen are not affected by this process.

A typical process flow diagram for an ion exchange ammonium removal system is shown in Figure 2-24. The system consists of a zeolite bed and a regenerant recovery unit. Regeneration is accomplished by either sodium chloride (neutral pH regeneration) or an alkaline reagent such as sodium or calcium hydroxide (high pH regeneration). High pH regeneration is more efficient than neutral pH regeneration. However, high pH regeneration may cause precipitation of magnesium hydroxide and calcium carbonate within the ion exchange bed. The most feasible regenerant recovery process has been air stripping of high pH regenerant.

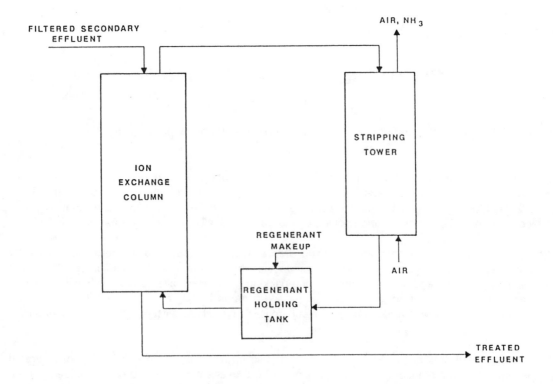

Figure 2-24. Schematic flow diagram for ion exchange nitrogen removal system.

2.7 References

1.   Metcalf and Eddy, Inc. **Wastewater Engineering: Treatment, Disposal, Reuse.** McGraw-Hill Book Company, New York, 1979.

2.   U. S. Environmental Protection Agency. **Process Design Manual for Nitrogen Control.** Washington, D.C., 1975.

3.   Won-Chong, G. M., and R. C. Loehr. The kinetics of microbial nitrification. **Water Research, 8,** 1099, 1975.

4.   Downing, A. L., H. A. Painter, and G. Knowels. Nitrification in the activated sludge process. **Jour. Inst. Sew. Purif., 32,** 130, 1964.

5.   Barnes, D., and P. J. Bliss. **Biological Control of Nitrogen in Wastewater Treatment.** E. & F. N. Spon, London, 1983.

6.   Hockenbury, M. R., and C. P. L. Grady. Inhibition of Nitrification -- Effects of Selected Organic Compounds. **Jour. Water Pollut. Control Fed., 49,** 768, 1977.

7.   Bridle, T. R., *et al.* Biological Nitrogen Control of Coke Plant Wastewaters. First Workshop on Water Pollution Control Technologies for the 80's, Wastewater Technology Centre, Burlington, Ontario, Canada, 1979.

8.   Anthonisen, A. C., *et al.* Inhibition of Nitrification by Ammonia and Nitrous Acid. **Jour. Water Pollut. Control Fed., 48,** 835, 1976.

9.   Beccari, M., R. Passino, R. Ramadori, and V. Tandoi. Kinetics of dissimilatory nitrate and nitrite reduction in suspended growth culture. **Jour. Water Pollut. Control Fed., 55,** 58, 1983.

10.   Barnard, J. Biological nutrient removal without the addition of chemicals. **Water Research, 9,** 485, 1975.

11.   Ekama, G. A., G. v. R. Marais, and I. P. Siebritz. Biological excess phosphorus removal. **Design and Operation of Nutrient Removal Activated Sludge Processes.** Water Research Commission, P.O. Box 824, Pretoria 0001, South Africa, 1984.

12.   Christensen, M. H. Denitrification of Sewage by Alternating Process Operation. **Progress in Water Technology,** Pergamon Press, **7,** 2, 339, 1975.

13.   Barnard, J. L. Cut P and N without chemicals. **Water and Wastes Engineering, 11**(7), 36 and **11**(8), 41, 1974.

14.   Argaman, Y. Design and performance charts for single-sludge nitrogen removal systems. **Water Research, 15,** 841, 1981.

15.   Henze, M., *et al.* Activated sludge model no. 1. IAWPRC Task Group on Mathematical Modeling for Design and Operation of Biological Wastewater Treatment. IAWPRC, 1987.

16.   Quirk, T. P., and W. W. Eckenfelder. Active Mass in Activated Sludge Analysis and Design. **Jour. Water Pollut. Control Fed., 58,** 932, 1986.

17.   Hui, A. M., *et al.* Pilot plant investigations of TF/SC and RBC/SC processes. Presented at the 55th WPCF Conference, Atlanta, Georgia, 1983.

18.   Gullicks, H. A., and J. L. Cleasby. Design of trickling filter nitrification towers. **Jour. Water Pollut. Control Fed., 58**, 60, 1986.

19.   Boller, M., and W. Gujer. Nitrification in tertiary trickling filters followed by deep-bed filters. **Water Research, 10**, 1363, 1986.

20.   Jiumm, M. H., C. Wu Yeun and A. Molof. Nitrified secondary treatment effluent by plastic media trickling filter.   Proceedings First International Conference on Fixed-Film Biological Processes, Kings Island, Ohio, 1982.

21.   Brenner, R. C., *et al.* Design Information on Rotating Biological Contactors. MERL, EPA, Cincinnati, Ohio, 1983.

22.   Tchobanoglous, G. Physical and chemical processes for nitrogen removal: theory and application. Proc. 12th Sanitary Engineering Conf., University of Illinois, Urbana, 1970.

23.   Folkman, Y., and A. W. Wachs. Nitrogen removal through ammonia release from ponds. Proceedings 6th International Conference on Water Pollution Research, Jerusalem, Israel, 1972.

Chapter 3

Design and Operation of

Biological Nitrogen Removal Facilities

3.1 Introduction

This chapter provides an overview on the design and operation of biological nitrogen removal facilities. Chapter 2 has already discussed the process fundamentals, including a review of the basic microbial processes, process stoichiometry and kinetics, and general process options. This chapter reviews specific options and compares them in order to assist the user in selecting the most appropriate option for a particular situation. Process and facility design are then reviewed, and process operation is discussed. Finally, the extent of use of biological nitrogen removal is discussed, and the performance capabilities of full-scale systems are described.

The use of single-sludge carbon oxidation/nitrification/denitrification systems is emphasized in this discussion due their cost-effectiveness and ease of use. However, the use of separate stage denitrification facilities is discussed, as appropriate.

Physical/chemical technologies for nitrogen removal are described in Chapter 2, but will not be addressed here. Experience with full-scale physical/chemical nitrogen removal facilities indicates several disadvantages relative to biological nitrogen removal facilities. Physical/chemical systems are generally more costly, more maintenance-intensive, and may have significant secondary environmental impacts (such as the atmospheric release of ammonia-nitrogen from a stripping process). For these reasons, biological nitrogen removal is generally the system of choice for most municipal applications.

Physical/chemical technologies are generally used only to polish the effluent from a biological nitrogen removal system. For example, breakpoint chlorination can serve as a back-up to a biological nitrogen removal system for those periods when operating upsets lead to less than complete nitrification. For detailed information on the design of physical/chemical nitrogen removal facilities, the reader is referred to the U.S. EPA **Process Design Manual for Nitrogen Control**(1) and the Water Pollution Control Federation Manual of Practice on **Nutrient Removal**(2).

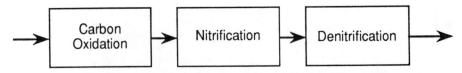

A. Separate Stage Carbon Oxidation, Nitrification, Denitrification

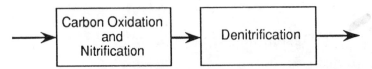

B. Combined Carbon Oxidation and Nitrification, Separate Stage Denitrification

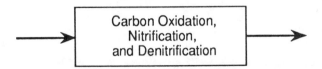

C. Combined Carbon Oxidation/Nitrification/Denitrification

Figure 3-1. Three major approaches to biological nitrogen removal.

3.2 Process Options

As discussed in Chapter 2 of this manual, the biological removal of nitrogenous compounds from typical municipal wastewater involves three basic processes:

o   Synthesis--incorporation of nitrogen into microbial mass as a result of cell growth

o   Nitrification--conversion of the ammonia and organic nitrogen commonly found in municipal wastewater to nitrate through oxidation by nitrifying microorganisms

o   Denitrification--conversion of the nitrate to nitrogen gas by denitrifying organisms, which is then released from the wastewater to the atmosphere

All of the various biological nitrogen removal systems currently available use these processes. The methods of incorporation of these processes into the treatment of municipal wastewater may be grouped into the following two basic categories based on the method of denitrification: (1) denitrification in a separate unit process, referred to as "separate stage denitrification"; and (2) combined carbon oxidation, nitrification, and denitrification, referred to as the "single-sludge" process. Nitrogen removal may be further categorized according to the approach used to accomplish nitrification. Figure 3-1 illustrates the three major approaches to biological nitrogen removal. Each approach is discussed in greater detail below.

3.2.1 Nitrification Options

As described above, nitrification (i.e., the biologically mediated conversion of ammonia and organic nitrogen to nitrate-nitrogen) is a necessary component to any biological nitrogen removal facility. Two general approaches are available to accomplish nitrification of municipal wastewater: (1) separate stage nitrification (Figure 3-1A), and (2) combined carbon oxidation and nitrification (Figures 3-1B and 3-1C). Separate stage nitrification involves the use of two biological processes in series. The first one removes carbonaceous biochemical oxygen demand (BOD) and the second one is used to nitrify the low-BOD effluent from the first process. In a combined carbon oxidation and nitrification system, the removal of BOD and nitrification are accomplished in a single biological process. Both nitrification approaches have been used successfully to nitrify municipal wastewaters. The choice between them depends primarily on cost factors.

3.2.2 Denitrification Options

3.2.2.1 Separate Stage Denitrification

Separate stage denitrification involves the use of a separate biological process to remove nitrate-nitrogen from the effluent of an upstream biological nitrification process. Either a separate stage nitrification system (Figure 3-1A) or a combined carbon oxidation and nitrification system (Figure 3-1B) may be used upstream of the separate stage denitrification system.

When separate stage nitrification is used with separate stage denitrification (Figure 3-1A), the overall biological nitrogen removal system consists of three biological processes operating in series and is referred to as a "three-stage" or "three-sludge" process. The first stage removes BOD, the second stage nitrifies the effluent from the first stage, and the third stage removes the nitrate-nitrogen contained in the effluent from the second stage. When a combined carbon oxidation and nitrification system is used with separate stage denitrification (Figure 3-1B), the overall biological nitrogen removal system consists of two biological processes operating in series and is referred to as a "two-stage" or "two-sludge" process. The first stage accomplishes BOD removal and nitrification, while the second stage denitrifies the nitrate-nitrogen contained in the effluent from the first stage. Therefore, in either a combined carbon oxidation/nitrification system or a separate stage nitrification system, denitrification in a separate stage denitrification system is accomplished in a separate unit process following carbonaceous BOD removal and nitrification.

Since carbonaceous BOD removal and nitrification leave wastewater largely devoid of readily available carbonaceous matter for denitrification, it is necessary to add an external carbon source to the wastewater. Methanol is typically used for this purpose. However, methanol addition must be carefully controlled to avoid adversely affecting the plant effluent BOD through overdosing.

Two different process options are typically used for separate stage denitrification: (1) suspended growth and (2) attached growth. These are described below.

3.2.2.1.1 Suspended Growth

This approach is analogous to an activated sludge treatment system. The wastewater first passes through a continuously mixed chamber, or reactor vessel, to which an external compound (in this case a carbon source, methanol) is added. The methanol is used as a carbon source by a group of microorganisms to accomplish the treatment objective. These microorganisms are then settled out in subsequent clarifiers and returned to the denitrification basin as return sludge. A portion of the sludge is removed, or wasted, from the system to maintain a desired mean cell residence time (MCRT), or solids retention time (SRT). However, unlike an activated sludge aeration basin the contents are not aerated. Instead, the contents are mixed with submerged devices to keep the biological solids in suspension while also maintaining the anoxic conditions necessary for denitrification. The reaction vessel is typically sized to provide an average detention time of 2 to 3 hours. An aerated channel or small aeration tank generally follows the denitrification reactor to strip the nitrogen gas bubbles from the microbial solids, thereby ensuring proper settling of the solids in the clarifiers. The aeration step may also be sized to oxidize any remaining methanol resulting from overdosing. Greater detail on the configuration of these systems is provided elsewhere(1,2).

3.2.2.1.2 Attached Growth

In this approach nitrified wastewater, to which an external carbon source (typically methanol) has been added, passes through one or more chambers, or vessels, which contain an inert media to which the denitrifying microorganisms are attached. Contact of the microorganisms with the wastewater is accomplished through distribution of the flow through the media, as opposed to the mechanical mixing required in the suspended growth system.

Several different types of attached-growth denitrification systems have been developed. These systems include the packed bed types, deep bed granular filtration types, and fluidized bed type. The packed bed system can be further separated into gas-filled and liquid-filled types.

The gas-filled packed bed system consists of a covered reactor filled with a plastic media (either in modules or random "dumped" type) through which the wastewater flows in a manner similar to a trickling filter. An atmosphere of nitrogen is maintained in the unit by virtue of the cover. Since a portion of the attached microorganisms continually slough from the media, a subsequent clarification or filtration step is needed.

The liquid-filled packed bed systems include both the high-porosity and low-porosity types. Both of these consist of enclosed chambers containing a media to which the denitrifying microorganisms are attached and maintained in a submerged state. The high-porosity type uses random dumped plastic media. A subsequent clarification process is needed to remove the microbial solids which continuously slough from the media. The low-porosity packed bed type uses a uniformly graded coarse sand as media. The unit serves the dual purpose of denitrification, through the attached growth on the media, and filtration. As such, periodic backwashing is necessary to remove accumulated solids and prevent blinding of the media.

A variation of this type of denitrification system is the deep bed granular filter. This filter consists of a relatively deep bed of coarse sand (6 feet) supported by a system of porous support plates and gravel, as in a typical gravity filter. Periodic backwashing is also required for this system to remove accumulated solids and trapped nitrogen gas. Figure 3-2 illustrates such a system.

The fluidized bed systems use either fine sand or activated carbon as media. The wastewater flow passes upward through the reactors and causes the media bed to expand. This expansion allows a greater microbial growth on the media particles without the accompanying problems of high head loss, channeling, and reduced efficiency which can occur with a packed bed system. Again, since microbial solids are continuously sloughed from the media, a subsequent clarification or filtration step is needed. Greater detail on the configuration of attached growth denitrification options is presented elsewhere(1,2).

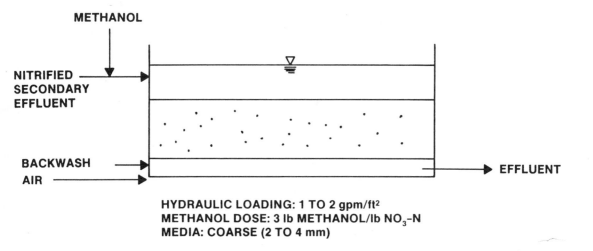

HYDRAULIC LOADING: 1 TO 2 gpm/ft²
METHANOL DOSE: 3 lb METHANOL/lb NO₃-N
MEDIA: COARSE (2 TO 4 mm)

Figure 3-2. Typical packed bed separate stage denitrification reactor.

3.2.2.2 Single-Sludge Denitrification

To avoid the operating costs associated with the continual addition of methanol required by the separate stage denitrification process, processes have been developed in which the carbon source naturally present in the wastewater is used to sustain denitrification (Figure 3-1C). These processes are referred to as "combined carbon oxidation/nitrification/denitrification" or "single-sludge." Two carbon sources are used in single-sludge biological nitrogen removal systems: (1) endogenous decay of the activated sludge microorganisms; and (2) the wastewater influent to the secondary treatment system. Either one or both of these carbon sources is used, depending on process configuration. Combined carbon oxidation and nitrification (as opposed to separate stage nitrification) is an inherent feature of these systems.

Systems using endogenous carbon sources were first suggested in the late 1960's and early 1970's. These systems simply add an anoxic reactor between the aeration basin and clarifier in a conventional nitrifying activated sludge system. The aeration system is designed to allow operation in the nitrification mode, and the resulting nitrate is denitrified in the anoxic basin. This system is easily incorporated into an existing activated sludge plant. However, it has the disadvantages of a very low denitrification rate due to the relatively low availability of carbon from endogenous decay and in the secondary effluent, and also the potential of some ammonia-nitrogen release due to the decay and lysis of biological solids.

In an attempt to minimize the ammonia release and the large anoxic reactor requirements of the low-rate endogenous carbon source system, treatment systems have been developed using the organic content of the influent wastewater for denitrification. Many process configurations have been suggested and evaluated. All of these include alternating aerobic/anoxic treatment zones or stages to increase the nitrogen removal capabilities of the process. The most typical of these systems is the four-stage Bardenpho process(1,3), shown schematically in Figure 3-3.

47

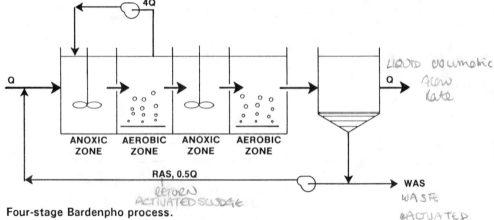

(handwritten annotations on figure: "LIQUID volumetric flow rate" near top right Q; "RETURN ACTIVATED SLUDGE" below RAS; "WASTE ACTIVATED SLUDGE" near WAS)

Figure 3-3. Four-stage Bardenpho process.

The four-stage Bardenpho system uses both wastewater carbon and endogenous decay carbon to achieve denitrification. The wastewater initially enters an anoxic denitrification zone to which nitrified mixed liquor is recycled from a subsequent combined carbon oxidation/nitrification zone. The carbon present in the wastewater is used to denitrify the recycled nitrate, which is then released as nitrogen gas in the aeration basin. The ammonia in the raw wastewater passes through the first anoxic zone unchanged to be nitrified in the first aeration zone. The nitrified mixed liquor which flows from the aeration zone then passes into a second anoxic zone, where additional denitrification occurs at a lower rate using the endogenous carbon source. A final period of aeration is provided prior to sedimentation to encourage release of the nitrogen gas and improve sludge settleability. Ammonia released from the sludge in the second anoxic zone is also nitrified in the last aerobic zone(4).

Other process approaches are used to achieve combined carbon oxidation/nitrification/denitrification. One example is through the use of an endless loop reactor, or oxidation ditch, shown schematically in Figure 3-4(5). In an oxidation ditch activated sludge system, mixed liquor flows continuously around a loop-type channel, driven and aerated by an aeration device located at one or more points in the channel. The aeration device may be a brush aerator, conventional low-speed aerator, submerged U-tube aerator, or any other device typically used in an oxidation ditch. Through the design and operation of the system, it is possible to create an aerobic zone capable of nitrification immediately downstream of the aerator, and an anoxic zone upstream of the aerator for some distance. By allowing the influent wastewater to enter the system at the upstream limit of the anoxic zone, some of the wastewater carbon source is used for denitrification. The effluent from the ditch is taken upstream of the anoxic zone and sent to a clarifier. This system, having only a single anoxic zone, is typically unable to achieve the same high nitrogen removals as the Bardenpho process(1).

Many other reactor configurations are possible. Systems have been constructed using only the first anoxic and first aerobic zone configurations of the Bardenpho system. They have proven to effectively remove nitrogen, although not to as low a level as achieved in the Bardenpho system. Sequencing batch reactor systems using anoxic and aerobic cycles can also be used to effectively simulate Bardenpho treatment sequences(6). Oxygen transfer systems can be operated intermittently in conservatively sized nitrification systems, resulting in the periodic creation of anoxic zones where denitrification will occur. Denitrification can also occur in a continuously aerated basin if a gradient in dissolved oxygen is created which allows a portion of the basin to remain anoxic(7). Importantly, all of these systems function according to the same basic principles and can be evaluated, designed, and operated according to those principles.

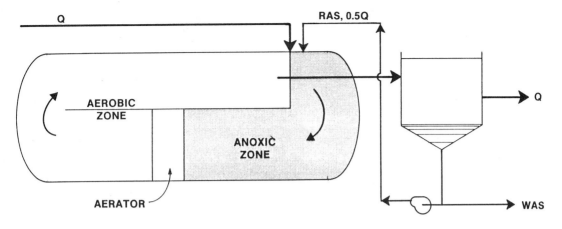

Figure 3-4. Looped reactor (oxidation ditch) configured for nitrogen removal.

3.3 Process Selection

Selection of a treatment process for nitrogen removal includes selection of an approach for both nitrification and denitrification. Carbon oxidation and nitrification are most often combined in a single process since the advantages of separating the two processes are rarely justified by the additional capital and operating costs of separate stage nitrification systems. However, two substantially different options are available for denitrification. Table 3-1 provides a qualitative comparison of the two denitrification approaches of "separate stage" and "single-sludge." In Table 3-1, the plus (+) sign indicates a favorable characteristic or feature of the particular option, and the minus (-) sign indicates an unfavorable characteristic or capability. A zero (0) indicates a neutral, or neither positive nor negative, characteristic.

Table 3-1. Denitrification process comparison.

		Separate Stage	Single-Sludge
Performance	Nitrogen removal	+	+
	TSS control	+ /-/0	0
Process Stoichiometry	Energy	-	+
	Alkalinity	-	+
	Carbon supplement	Required	None (Internal)
Operation and Maintenance	Control	+	+
	Operations	0	+
	Chemical storage and handling	-	+
	Maintenance	0	0
Cost	Capital	Equal to Higher	Equal to Lower
	O&M	Higher	Lower

49

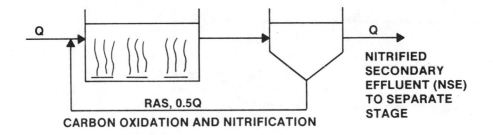

CARBON OXIDATION AND NITRIFICATION

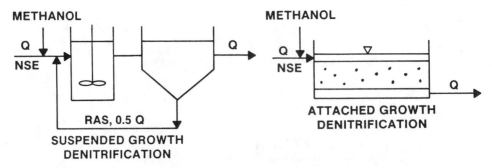

A. SEPARATE STAGE NITROGEN REMOVAL SYSTEMS

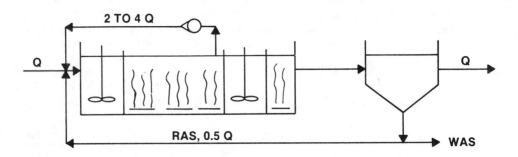

B. SINGLE SLUDGE NITROGEN REMOVAL SYSTEM

Figure 3-5. Nitrogen removal systems considered in comparison.

Figure 3-5 presents schematics of the two denitrification approaches. As discussed above, separate stage systems receive a nitrified secondary effluent (NSE) produced in an upstream nitrification system, here represented by a combined carbon oxidation and nitrification system. Consequently, addition of an external carbon source, such as methanol, is required. Figure 3-5A depicts a combined carbon oxidation and nitrification system treating an influent wastewater (raw sewage or primary effluent) to produce a NSE. Methanol is then added to the NSE as a carbon source for denitrification in either a suspended growth or attached growth system. By comparison, a single-sludge system (Figure 3-5B) receives influent wastewater (raw sewage or primary effluent) and uses the carbon contained in the influent for denitrification. The biological reactor consists of aerobic zones for nitrification and anoxic zones for denitrification.

3.3.1 Performance

The separate stage and single-sludge denitrification processes can both achieve high removals of nitrogen, on the order of 85 to 95 percent. As will be discussed in greater detail below, similar quality effluent can be achieved by both processes. The single-sludge process (Figure 3-5B) does not enhance or degrade control of total suspended solids (TSS) in the effluent from the process, behaving very similar to a comparable nitrifying activated sludge process. However, the separate stage process may either impede or enhance the control of TSS in the effluent, depending on the type of separate stage denitrification process used. The filter-type attached growth systems (Figure 3-2) can have a beneficial impact on effluent TSS levels due to filtration. However, the attached growth systems which continuously slough microbial solids can result in increased TSS levels in the effluent due to poor settleability of those solids. Suspended growth separate stage denitrification systems have a neutral effect on effluent TSS levels.

3.3.2 Process Stoichiometry

The primary difference between the separate and single-sludge denitrification systems is the source of carbonaceous material which serves as the electron donor in the denitrification reaction. As described previously, the naturally occurring carbonaceous compounds in the wastewater serve as the carbon source in the single-sludge denitrification process. In contrast, a separate compound, typically methanol, must be added to provide a carbon source for the separate stage process. Therefore, the single-sludge process is a self-contained process with respect to the stoichiometry of denitrification, while the separate stage system requires the input of external biochemical energy. In addition, since denitrification supplies the equivalent of 2.86 pounds of oxygen per pound of NO_3-N removed, the initial denitrification step in the single-sludge system actually reduces the energy requirements for BOD removal from the wastewater. On the other hand, the BOD is removed prior to denitrification in a separate stage system, and the BOD-oxidizing characteristic of denitrification is not a benefit to the overall process.

Another consideration in nitrification/denitrification systems is the balance of alkalinity in the system. As discussed in Chapter 2, nitrification *consumes* alkalinity (7.2 pounds as $CaCO_3$ per pound of NO_3-N generated) while denitrification *produces* alkalinity (3.6 pounds as $CaCO_3$ per pound of NO_3-N removed). In a single-stage process, the alkalinity is produced by denitrification *prior* to the alkalinity consuming nitrification process. Approximately one-half of the alkalinity required by the nitrification process is produced in the preceding anoxic basin. On the other hand, in the separate stage process the denitrification step *follows* the nitrification step. While the net effect on the plant effluent is similar, a low alkalinity wastewater could experience a pH drop in the aeration basins that would require offsetting chemical addition, if a separate stage system were used.

3.3.3 Operation and Maintenance

Each of the two approaches to denitrification has its own unique operational considerations. Both systems require similar control of the carbon oxidation/nitrification step to ensure adequate nitrification over varying flows, loadings, and wastewater temperatures. For the single-sludge systems, the denitrification process is controlled by the rate of nitrate recycle (in the mixed liquor) to the first anoxic zone. The return of nitrate in the return activated sludge (RAS) and the recycled mixed liquor (ML) controls the mass of nitrate directed to the first anoxic zone, thus establishing an upper limit on nitrate removal in that zone. The quantity of organic matter present in sewage can also be limiting, particularly if certain high-nitrogen industrial wastewaters are also being treated. High dissolved oxygen concentrations in the recycle mixed liquor can impact denitrification capabilities for weak influent wastewaters.

The primary operation controlling performance in separate stage systems is the rate of methanol addition. An aerated chamber or channel is typically provided following the denitrification step in suspended growth systems to oxidize any remaining methanol. However, excessive overdosing of methanol could exceed the capacity of the supplemental aeration step and result in an increase in the effluent BOD. Aeration following an attached growth system is not effective in oxidizing methanol, thereby increasing the need for careful operation of these systems(1). Thus, while an equivalent degree of control over effluent quality is available in both single-sludge and separate stage systems, operation is more difficult for the separate stage system. In addition, the single-sludge system does not require the use of external chemicals, while the separate stage system involves the storage and handling of methanol. Methanol is flammable, explosive, and hazardous to breathe. Consequently, special procedures are required for its safe storage and handling.

Neither the separate stage nor single-sludge processes have particular drawbacks with respect to maintenance. Some additional maintenance may be expected for the filter-type attached growth systems to replace media in the event of excessive fouling.

3.3.4 Cost

3.3.4.1 Capital

The initial and second stage reactors in single-sludge nitrogen removal systems require either a separate basin or portions of a common basin structure. The use of common basin construction is preferable as it is less costly. In addition, separate mixed liquor (nitrate) recycle pumping must be provided. For suspended growth separate stage denitrification, a smaller anoxic reactor is required than for a single-sludge system. However, another set of clarification and return sludge pumping facilities must be constructed. This results in a capital cost that often exceeds that of comparable single-sludge systems. Attached growth denitrification systems require a structure to contain the media, an underdrain system, and a backwash system. Costs must be developed specifically, but often exceed those of a single-sludge system. Both of the separate stage system types also have the added cost for methanol storage and feed equipment. As a result of the above factors, the separate stage denitrification system will typically have a higher initial capital cost than will the single-sludge systems.

3.3.4.2 Operation

Two major operational cost items for nitrogen removal systems are electrical power and chemicals. For the single-sludge systems, the recycle of mixed liquor to the anoxic basin and the mixing of the larger volume first stage and second stage anoxic zones all consume additional power compared to a separate stage system. However, these costs are typically offset by the removal of some BOD and the elimination of the associated aeration requirements through denitrification in the single-sludge process. For the separate stage denitrification processes, the primary additional operating cost (compared to the single-sludge process) is for the methanol. In addition, there could be added cost for chemical addition to control pH in a low alkalinity wastewater resulting from the nitrification process, as discussed previously. Operating labor costs may also be greater for separate stage systems since more unit processes must be operated. Due primarily to the cost of the methanol, separate stage denitrification systems generally have higher operating costs than do single-sludge systems.

3.3.5 Summary

The previous discussion suggests that single-sludge biological nitrogen removal systems will often be the system of choice for most municipal wastewater treatment applications. These systems are generally the most cost-effective and the most desirable from an operational standpoint. They have the added advantage of using technology familiar to operators of typical activated sludge systems (i.e., pumps, mixers, aerators, etc.). A separate stage system might be more desirable when effluent filtration is necessary to meet a stringent effluent suspended solids criteria. In this case, the low-porosity type packed bed reactor, or filter, may be more cost-effective than providing a separate filter. Consideration should also be given to the need for phosphorous removal. If phosphorus removal is needed, the selection of biological phosphorus removal will directly impact the type of nitrogen removal process selected, as discussed in Chapters 6 and 7.

Because of their popularity, sections 3.4 and 3.5 of this chapter will focus on single-sludge systems for combined carbon oxidation, nitrification, and denitrification. However, separate stage denitrification systems will be included in section 3.6 on full-scale experiences. Note that attached growth, separate stage denitrification systems are often proprietary and detailed design and performance data can be obtained from the vendors of these systems.

3.4 System Design--Single-Sludge Systems

This section discusses the design of single-sludge systems for carbon oxidation, nitrification, and denitrification. Topics considered include process design, facility design, and facility costs. Such systems are essentially modifications to conventional nitrifying activated sludge facilities to incorporate anoxic zones and mixed liquor recycle pumping. The basic process was described previously and is illustrated in Figure 3-3. Other process configurations incorporating only the first or second anoxic zone are possible when less than complete nitrogen removal is required. The procedures described in this section are equally applicable to the design of such modified systems.

The presentation in this section assumes that the reader is familiar with the design and operation of conventional nitrifying activated sludge systems. Consequently, differences or additions are described, rather than basic process components. The reader is referred to Chapter 2 for a review of basic principles, as well as to standard texts on activated sludge design and operation for other relevant background information.

3.4.1 Process Design

Single sludge nitrogen removal systems incorporate three processes: (1) carbon oxidation, (2) nitrification, and (3) denitrification. In these systems, carbon oxidation and nitrification are accomplished in the aerobic zones. Detailed procedures and calculations for system design for carbon oxidation and nitrification are presented elsewhere(1,2,8). Denitrification occurs in the anoxic zone. Detailed procedures and calculations for design of the denitrification process are presented in Chapter 6 of **Theory, Design, and Operation of Nutrient Removal Activated Sludge Processes** published by the Water Research Commission, Republic of South Africa(4). A detailed model for the design and evaluation of such processes has recently been published(9,10,11).

The discussion presented here provides an overview of nitrogen removal process design. It is not intended to serve as a detailed step-by-step design guide; the reader is referred to the above referenced publications for detailed design procedures and calculations. However, the simplified procedure discussed below illustrates the conceptual basis for process sizing and provides an approach for checking process designs developed using more sophisticated approaches. The procedure consists of the following major elements, which are briefly considered in the subsequent sections of this chapter:

1. Sizing of first aerobic zone
2. Sizing of anoxic zones and mixed liquor recycle pumping
3. Sizing of second aerobic zone, if needed
4. Clarifier sizing
5. Overall process mass balance check

3.4.1.1 First Aerobic Zone

Nitrogen must be converted to nitrate in a single-sludge biological nitrogen removal system before it can be removed through denitrification. Since nitrification is an aerobic process, it can occur only in the aerobic zone. Consequently, it is logical to begin the design of a single-sludge system with the sizing of the first aerobic zone.

As explained in Chapter 2, nitrification can be achieved when the solids retention time (SRT), or mean cell residence time (MCRT) for solids within the first aerobic zone exceeds some critical value representing the maximum growth rate of the nitrifying bacteria. As illustrated in Figure 3-6 (developed using the procedures outlined in reference 1 assuming typical operating characteristics), the minimum aerobic SRT is affected significantly by temperature. Consequently, the design minimum aerobic SRT is based on the lowest sustained operating temperature for the biological system. Weekly or monthly average minimum biological reactor temperatures, as opposed to minimum daily temperatures, are typically selected for this purpose. For example, consider a system where the minimum weekly average biological reactor temperature is estimated to be 10°C. At 10°C, the minimum aerobic SRT would be about 5.5 days under the conditions used to generate Figure 3-6. The minimum aerobic SRT is then multiplied by a safety factor, typically between 1.5 and 2.5, to obtain the design aerobic SRT. As described in Chapter 2, the safety factor is necessary to provide for stable operation and acceptable effluent quality with varying influent conditions. For operation at 10°C, a safety factor of 2.0 to 2.5 would typically be used, resulting in a design value between 11 and 14 days.

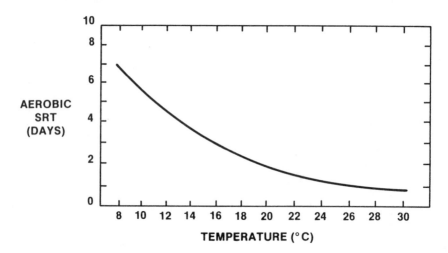

Figure 3-6. Effect of temperature on the minimum SRT for nitrification(pH, 7.2; DO, 2 mg/L).

Sizing of the aerobic zone also depends on the process sludge yield and the design mixed liquor suspended solids (MLSS) concentration. Procedures for calculating process sludge yields are described elsewhere(1,2,4,8,12). However, values often range between 0.6 and 0.8 lb TSS/lb BOD applied when primary effluent is being treated and between 0.8 and 1.0 lb TSS/lb BOD applied for raw sewage. Design MLSS concentrations often range between 2,500 and 3,500 mg/L, as constrained by the secondary clarifier solids loading rate.

The approach described here is essentially identical to that which would be used to size the biological reactor for a combined carbon oxidation and nitrification system. Thus, the size of the first aerobic zone of a single sludge nitrification/denitrification system is identical to that of a combined carbon oxidation and nitrification system. System performance will be similar in both cases in terms of ammonia-nitrogen removal. Effluent ammonia-nitrogen concentrations will typically be in the 0.5 to 2 mg N/L range during periods of stable operation.

In summary, sizing of the first aerobic zone consists of the following steps:

1. Select design aerobic SRT based on lowest anticipated monthly or weekly average operating temperature and selected factor of safety.

2. Calculate secondary sludge production based on process BOD loading and yield.

3. Multiply secondary sludge production (e.g., lb/day) by design aerobic SRT (days) to obtain required first aerobic zone solids inventory.

4. Convert required inventory into tank volume based on design MLSS concentration.

3.4.1.2 Anoxic Zones

The process design of the anoxic zones includes determination of the anoxic reactor volumes and the required mixed liquor recycle rate. Appropriate values may be determined using equations such as those presented in Chapter 2 and in the references noted above(4). The IAWPRC model, when calibrated to a particular wastewater, may also be used(11). An alternative, simplified procedure is discussed below.

It is necessary to balance the fraction of sludge held under aerobic conditions for nitrification against the fraction of sludge which is not aerated and available for denitrification. The non-aerated fraction must be further subdivided between the first and second anoxic zones. In fact, when only 60 to 75 percent nitrogen removal is required, the second anoxic zone may be omitted entirely. The denitrification potential of the primary anoxic zone is directly dependent on the minimum wastewater temperature and the influent biodegradable chemical oxygen demand (COD). The best denitrification performance is achieved when the nitrate loading on the first anoxic zone, which is controlled through the mixed liquor recycle rate, is equal to its denitrification potential. The sizing of the anoxic basin is dependent on the assumed mixed liquor concentration and the non-aerated sludge mass fraction in each of the anoxic zones.

The first anoxic zone is sized to remove the nitrate delivered to it in the mixed liquor recycle based on specific rates of denitrification reported in the literature. The second anoxic zone is then sized to remove any remaining nitrate, as necessary to meet the discharge permit. The procedure is as follows. First calculate the nitrogen that will be nitrified in the process. This can be calculated as the influent total kjeldahl nitrogen (TKN) minus nitrogen uptake by the secondary sludge. The nitrogen content of secondary sludge is typically 8 to 12 percent of the volatile solids content. Thus, the nitrogen to be nitrified is the influent mass of TKN minus the quantity of secondary sludge production times the nitrogen content of the secondary sludge.

Next, the mixed liquor recycle rate is calculated. Basically, it must recycle the quantity of nitrate removed in the first anoxic zone. The quantity of nitrate removed in the first anoxic zone is the nitrate generated in the first aerobic zone minus the quantity of nitrate directed to the second anoxic zone. By selecting the nitrate concentration of the first aerobic zone, the quantity of nitrate reduction in the first and second anoxic zones can be calculated. Using the quantity of nitrate to be reduced in the first anoxic zone, the mixed liquor recycle rate is calculated. Typically, the first anoxic zone and mixed liquor recycle is sized to remove 65 to 85 percent of the nitrate which must be removed. The first aerobic zone nitrate concentration is 4 to 8 mg/L as nitrogen, and recycle pumping capacities are between 100 and 400 percent of the plant influent flow rate.

Both the first and second anoxic zones can then be sized based on appropriate specific rates of denitrification. The rate will be higher in the first anoxic zone than in the second due to the greater availability of raw wastewater COD. Correlations between specific rates of denitrification and other operating parameters have been summarized in Chapter 2. For example, Burdick *et al.*(3) indicate that the specific rate of denitrification in the first anoxic zone is related to the food to microorganism loading ratio (F/M) loading on the first anoxic zone as follows:

$$SRDN_1 \quad = \quad 0.03 \ (F/M_1) \ + \ 0.029 \qquad\qquad (1)$$

where: $SRDN_1$ =   specific rate of denitrification in the first anoxic zone, g NO_x-N/g MLSS-day

F/M_1 =   F/M loading ratio on the first anoxic zone, g BOD/g MLSS-day

Specific rates of denitrification in the first anoxic zone typically range from 0.05 to 0.15 g NO_x-N/g MLSS-day and depend primarily on the organic loading rate on the anoxic zone (F/M_1) and the nature of the wastewater. Rates, as calculated using equation (1), may be considered to be typical for an average municipal wastewater. Specific rates of denitrification in the second anoxic zone are typically 20 to 50 percent of the rate in the first anoxic zone. Burdick *et al.*(3) present a correlation between the overall process SRT and the specific rate of denitrification in the second anoxic zone. In either case, from the mass of nitrate to be removed in each zone, the appropriate specific rate of denitrification, and the design MLSS concentration, the volumes of each zone can be calculated. A design example using this approach is presented in EPA **Design Manual for Phosphorus Removal**(13).

In summary, the procedure for sizing the anoxic zone(s) consists of the following steps:

1. Calculate nitrogen to be nitrified.

2. Select first aerobic zone effluent nitrate concentration. Based on this value calculate nitrate to be denitrified in the first anoxic zone and the required mixed liquor recycle rate.

3. Select appropriate specific rates of denitrification for the first and second anoxic zones.

4. Based on the quantity of nitrate to be removed in each zone, the specific rates of denitrification, and the design MLSS concentration, calculate the size of each anoxic zone.

Steps 2 through 4 can be repeated using different first aerobic zone effluent nitrate concentrations to optimize overall system sizing and removal.

3.4.1.3 Second Aerobic Zone

In the single-sludge system, the second aerobic zone serves the following two purposes: (1) strip nitrogen gas produced in the secondary anoxic zone from the microbial solids to ensure proper settling in the subsequent clarifier(s); and (2) nitrify any ammonia produced in the second anoxic zone due to endogenous decay. Consequently, it is needed only if a second anoxic zone is provided. The design of this zone is relatively simple, involving sizing of the zone and determination of the aeration requirements. The zone should be sized to provide 30 to 45 minutes theoretical hydraulic residence time at average flow. The aeration requirement should be determined based on the anticipated nitrification requirement for the ammonia leaving the secondary anoxic zone, and the endogenous carbonaceous oxygen demand of the mixed liquor. This oxygen required must be compared against the energy required for mixing the chamber contents (about 20 cfm/1,000 ft^3 for diffused air, or 0.6 to 1.15 hp/1,000 ft^3 for mechanical surface aeration). The larger of the two requirements (aeration/mixing) should be provided.

3.4.1.4 Secondary Clarification

The process design of the secondary clarifiers is essentially the same as that for a typical activated sludge system, with the numbers and sizes of clarifiers as necessary to provide a surface overflow rate (at average flow) of approximately 300 to 600 gallons per square foot per day (gal/ft^2-day). Secondary clarifier solids loading rates should also be reviewed. High design average overflow rates are discouraged since some single-sludge facilities have exhibited a tendency toward sludge bulking under certain conditions.

3.4.1.5 Mass Balance Checks

After the individual process components have been sized, the design should be checked by calculating various mass balances. The total process oxygen mass balance should be determined, based on the carbonaceous/nitrification oxygen requirements as offset by the credit resulting from consumption of BOD during denitrification. A mass balance on alkalinity should also be calculated, including both the alkalinity produced in denitrification and that consumed in nitrification. This calculation will indicate whether pH adjustment by chemical addition will be necessary prior to the first aerobic zone. Finally, the overall total process SRT should be calculated to make sure that it is within a reasonable range.

3.4.2 Facilities Design

Proper detailed design of facilities required to implement biological nitrogen removal is critical to the successful operation and performance of a system. Important facility design issues are discussed in this section.

As noted previously, single-sludge systems typically use common basin construction with the aerobic and anoxic zones located in various portions of the same basin structure, separated by walls and gates. This reduces construction cost through common wall construction and reduced structural requirements since many dividing walls need not be water holding. It also conserves plant site, which is often critical when expanding an existing plant on a constrained site. The arrangement of the zones and interconnecting gates and channels should be such that operational flexibility and component by-passing can be achieved. Flexibility and operability requirements must be judged for each application.

3.4.2.1 First Aerobic Zone

The design of the first aerobic zone is similar to that for a typical activated sludge system. In fact, it may be viewed simply as the aeration basin for a nitrifying activated sludge system. The system may be designed as either a plug flow or complete mix basin configuration. Improved performance will result from a plug flow configuration. However, the higher oxygen requirements of nitrification can create loading problems at the head end of a plug flow system, and this factor should be considered in designing the aeration system. Three different classes of aeration equipment are typically used in activated sludge aeration systems: (1) mechanical surface aerators; (2) fine or coarse bubble diffused air systems; and (3) submerged turbine aerators. These systems each have different associated oxygen transfer efficiencies, although other operational and maintenance characteristics often override the efficiency factors.

Although they require little maintenance, mechanical surface aerators may not be the system of choice for nitrification due to their limited turndown capability and high heat loss in cold weather applications. This is a disadvantage, since wide variations in oxygen requirements of the process result from diurnal and seasonal changes. If the aeration system can be adjusted to more closely match those varying needs, the opportunity exists for energy savings.

Diffused air systems are well-suited to nitrification systems since they have a much wider turndown range. It is also easier to provide tapered aeration for a plug flow configuration with a diffused air system than with mechanical surface aerators. Due to the relatively high aeration requirements for nitrification, fine bubble diffused aeration (with its higher oxygen transfer efficiency) is preferable over coarse bubble diffusion. However, this higher efficiency comes with a potential for greater maintenance due to diffuser fouling.

Submerged turbine aerators have the advantages of diffused air in terms of turndown capability, although the energy drawn by the mixer portion of the aerator is essentially fixed with the turndown savings being in the air flow to the diffuser. This type of aerator has the additional advantage of being easily converted to a mixer by simply shutting off the air flow. This can provide additional system flexibility in a plug flow basin configuration by allowing adjustment of the aerobic and anoxic zones.

For any aeration system a dissolved oxygen monitoring/aeration control system should be considered. The savings in aeration energy resulting from turndown during diurnal periods of low demand can be substantial and can easily offset the additional capital cost of the control system. Typically the system would consist of one or more in-situ dissolved oxygen sensors coupled to a control system which either adjusts air delivery (for diffused air systems and submerged turbine aerators) or basin level/aerator speed (for mechanical surface aerators). A small programmable controller is well suited for this control scheme as it can allow more complex time- and dissolved oxygen-related control decisions than typical hardware-based logic systems using relays, timers, and analog controllers. The primary drawback to this type of control system is that the dissolved oxygen monitoring device, which is central to the control system, requires significant attention in terms of maintenance and calibration to ensure a representative measurement of basin conditions.

3.4.2.2 Second Aerobic Zone

The design of the second aerobic zone is generally much simpler than that for the primary aerobic zone. The oxygen requirements for this zone are relatively low and are also relatively constant. The turndown concerns typical of the primary zone are, therefore, not significant for the secondary zone. As a result, dissolved oxygen control systems are not justified for this zone. Typically the same type of aeration equipment is used for both the first and second aerobic zones.

3.4.2.3 Anoxic Zones

The anoxic zones have two basic required features: (1) a basin or walled-off segment of a basin of sufficient volume; and (2) sufficient mixing of the contents to maintain the microbial solids in suspension without transferring oxygen to the contents. Submerged propeller or turbine mixers are typically used for this latter purpose. Figures 3-7 and 3-8 illustrate each type of mixer.

These devices mix without breaking the water surface, as does a mechanical surface aerator. They are capable of maintaining biological solids in suspension at minimal energy inputs. While energy input is an important variable in obtaining solids suspension, the number and placement of the mixers is more important. Consequently, the manufacturer should be contacted for specific installation details. Propeller mixers tend to work similar to a fan, with a spreading plume of mixing energy emanating from the mixing device. If not sufficient in number or properly oriented with respect to the basin configuration, it is possible to have localized dead spots which will become anaerobic and cease to denitrify. The mixer manufacturer should be consulted in designing the mixer layout.

Baffles used to define the anoxic zones also should be designed to allow floating solids to exit the system. Designs which trap floating solids can result in significant accumulations of scum, leading to odor and other operating problems. The use of submerged baffles, as illustrated in Figure 3-8, is encouraged. In such a design, floating solids can pass from one zone to another, finally exiting the aeration basin where they can be collected in the secondary clarifiers. Collected solids should be wasted to the solids handling system, not recycled to the head of the treatment plant. These simple design details can significantly reduce the accumulation of floating solids, and the associated problems.

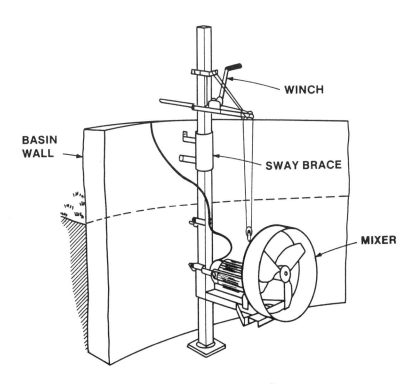

Figure 3-7. Typical propeller mixer.

59

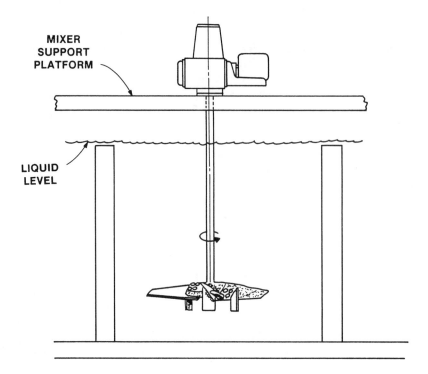

MIXER
SUPPORT
PLATFORM

LIQUID
LEVEL

Figure 3-8. Typical submerged turbine mixer.

3.4.2.4 Recycle Pumping

The recycle of mixed liquor from the first aerobic zone to the first anoxic zone is generally accomplished by pumping. Since the water level in the two zones is virtually the same, the only pumping head is due to pipe friction and fitting losses. However, offsetting the low head requirement is the high pumping volume required. The typical recycle ratio (with respect to plant flow) ranges from 1:1 to 4:1, but ratios as high as 6:1 may be required in some cases, particularly with a higher strength sewage.

Rather than constructing a separate dry-pit pumping facility, low-head submersible non-clog sewage pumps, propeller pumps, or non-clog vertical turbine pumps are generally mounted directly in the aerobic basin. The pumps should be located near the downstream end of a plug flow aerobic chamber. Regardless of the type of aerobic zone, however, the pumps should not be located immediately adjacent to an aeration device. Using this approach, the amount of dissolved oxygen (DO) returned with the mixed liquor will be minimized. Flow should be conveyed in a pipe rather than a channel to avoid DO entrainment. The discharge to the anoxic zone should be submerged for the same reason.

Another consideration in the design of the recycle pumping facilities is the variation of the recycle pumping rate. Unlike a raw sewage pump station at a plant, where it is necessary to match a varying influent flow rate, the mixed liquor recycle flow must only be within a specified range based on a specific plant flow. Hence, there is no need to specifically match the varying plant flow, and constant speed pumps may be used. However, to accommodate seasonal variations in nitrogen loading and wastewater temperature, it is desirable to have a sufficient number of recycle pumps so that the flow can be varied step-wise to optimize the process and/or avoid excess energy usage.

3.4.2.5 Secondary Clarification

Some nutrient removal systems have a tendency to develop a troublesome scum which can cause odor problems and degradation of the plant effluent quality. As discussed above, the basin should be designed to allow floating solids to pass to the secondary clarifiers. Consequently, design of the secondary clarifier scum removal and handling facilities to deal effectively with the potential of excessive scum development is prudent for a nitrogen removal plant.

The clarifier mechanism should include a positive means of scum removal. One example is the full radius "ducking skimmer/rotating weir" arrangement. As illustrated in Figure 3-9, this device includes a pipe with a slot cut along the centerline on one side to serve as a weir. As the full radius scum skimmer sweeps toward the pipe, the pipe rotates downward and a scum/water mixture flows over the weir edge and into the pipe. This mixture then flows to one end of the pipe where it is discharged to the scum pumping facilities. The rotating weir pipe should extend into the feed well to remove scum from that area also. A sprayer should be directed at the upstream end of the weir pipe and operated automatically to assist in scum flow down the pipe. If properly designed, this type of scum removal mechanism will perform significantly better than the standard 4-foot-wide scum trough and beach at the clarifier perimeter.

The full radius design represents one of several high-volume secondary scum collection devices now available and illustrates the importance of secondary scum collection. Again, it is emphasized that collected scum must be wasted from the system, not recycled within the liquid process train. Collection and wastage of floating solids are key to minimizing scum and foaming problems in biological nitrogen removal facilities.

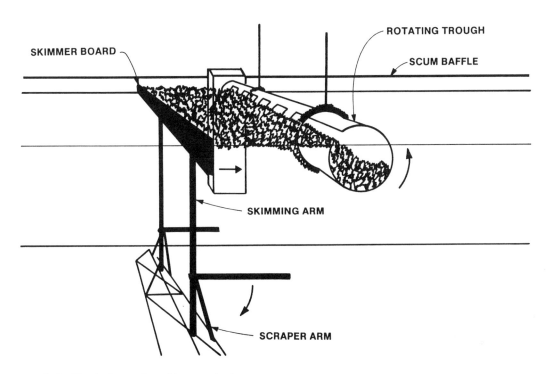

Figure 3-9. Typical rotating skimmer device.

3.4.3 Facilities Costs

The incorporation of biological nitrogen removal into a new or existing secondary wastewater treatment plant requires facilities and equipment that would otherwise not be necessary. These added facilities translate into added treatment costs for the removal of nitrogen. In general, the size of any additional facilities required for single-sludge biological nitrogen removal is determined primarily by the pollutant (BOD and TKN) loading on the process. Their size (and cost) is relatively independent of plant flow. Consequently, the development of standardized "cost curves" based on plant flow is not possible for these systems.

The following paragraphs discuss, and in some cases quantify, the approximate additional capital costs associated with single-sludge biological nitrogen removal facilities. This discussion is in no way intended to replace a detailed, plant specific equipment and unit process sizing exercise, with an associated construction cost estimate prepared by a professional estimator. Rather, the intent is to indicate the various facilities and equipment for which costs must be generated, and to provide an order-of-magnitude measure of the probable capital costs for these facilities.

The emphasis here is on the added cost for nitrogen removal over and above that which would be required for a standard activated sludge system. Other resources are available to develop preliminary order-of-magnitude cost estimates for a wide variety of nitrogen removal options. Of particular interest is the information presented in the U.S. EPA **Innovative and Alternative Technology Assessment Manual**(14) and the cost curves presented in the Water Pollution Control Federation Manual of Practice on **Nutrient Control**(2). The material presented in this section should be considered supplementary to these two references.

Basin-wide evaluations of the costs to retrofit nitrogen removal to municipal wastewater treatment plants illustrate the site-specific nature of retrofit costs. In two recent studies conducted in the Chesapeake Bay region which considered the individual characteristics of a large number of facilities (15,16), incremental capital costs averaged approximately $1 per gallon per day (gpd) of capacity. However, incremental costs ranged from under $0.1 to over $4 per gpd of capacity. Many plants were in the $0.6 to 0.8 per gpd capacity range. Total incremental treatment costs (amortized capital plus operation and maintenance) averaged approximately $0.6/1,000 gallons treated. However, the range was from less than $0.1 to over $1 per 1,000 gallons treated. Incremental costs will vary depending on wastewater characteristics, the nature and condition of the existing facilities, site constraints, and numerous other factors. These results emphasize that site-specific evaluations must be conducted if cost impacts are to be assessed realistically.

3.4.3.1 Basins

One of the primary requirements for nitrogen removal is the provision of basin volume in which the nitrification and denitrification reactions can occur. As discussed previously, nitrification will increase the aeration volume required, and denitrification will require additional basin volumes. At a new plant, these volumes can be provided either by separate basin structures or by a single structure employing common wall construction. At an existing plant, the reactor volumes can be provided by construction of additional basins, or by utilizing existing structures. The latter can be accomplished most easily by either converting an existing structure from another use no longer needed, or by apportioning excess volume in an existing aeration basin through the addition of baffle walls to provide a separate anoxic basin.

To provide an indication of the added costs to provide nitrification and denitrification, in addition to carbon oxidation, construction costs have been estimated for a series of concrete aeration basins ranging from 0.5 million gallons(MG) to 2 MG. The basins were assumed to be 15 feet deep plus 3 feet freeboard, and constructed with the top of the basin roughly at grade. These estimates indicate a unit cost ranging from $0.75 per gallon of working volume for the 0.5 MG basin, down to $0.50 per gallon of working volume for the 2 MG basin, or an average of $0.60 per gallon. The costs include allowances for finishes, miscellaneous metal work, and other non-quantified items, as well as contractors' mobilization and general/administrative overheads. However, they are only an indication of the order of magnitude of cost for buried basins. There are many site-specific design and construction factors that could significantly alter the actual costs, generally resulting in increased unit and total construction costs.

If adequate existing basin volume exists to allow an apportionment of the volume through the addition of concrete baffle walls, a cost for the baffle walls may also be estimated. The cost may be estimated on a square-foot basis. It is, of course, directly dependent on the wall thickness and, to a lesser extent, the wall height. Since the baffle walls are generally not designed to withhold water, an 8-inch to 12-inch thick wall will typically be adequate. An approximate cost for a 15-foot high baffle wall will range from $7.50/ft^2 for an 8-inch reinforced concrete wall to $11.00/ft^2 for a 12-inch wall.

3.4.3.2 Aeration Systems

As discussed previously, biological nitrogen removal requires significant additional aeration capacity for the nitrification process as compared to carbon oxidation, and additional aeration equipment for a secondary aerobic zone if two-stage denitrification is used. If mechanical surface aeration is used, this translates into larger horsepower aerators and possibly a greater number of them. For a diffused air system, the added aeration requirements translate into a greater number of diffusers, more piping, and larger capacity (and possibly a greater number of) blowers. For a submerged turbine aeration system, the need for additional aeration requires that the blower sizes and/or numbers be increased, and that the number of submerged turbine aerators also be increased. The costs for aeration systems may be estimated based on the required oxygen input, expressed in pounds per day (lb O_2/day), and the efficiency of the aeration system in pounds per horsepower per hour (lb O_2/hp-hr).

For surface mechanical aeration systems, the added cost for nitrification may be estimated by first determining the oxygen requirements for carbonaceous oxidation only versus that for carbonaceous oxidation plus nitrification as explained earlier in this chapter. The required aerator horsepower may then be determined based on an assumed efficiency, for example 1.5 lb O_2/hp-hr. From the total horsepower requirement and approximate basin configuration, the number of aerators and horsepower per aerator should then be determined. The installed cost of mechanical aeration equipment varies from $1,900 per horsepower for 25 horsepower aerators to $750 per horsepower for 100 horsepower aerators. For an efficiency of 1.5 lb O_2/hp-hr, this translates to a range of $20 to $53 per lb O_2/day of capacity for mechanical equipment.

A diffused air aeration system includes blowers and a diffuser/piping system in the aeration basin. The costs for such a system are greatly influenced by the type of diffusers, their transfer efficiency, and arrangement in the basin. As such, the diffused air equipment manufacturer should be consulted for detailed cost estimated for this type of system. However, an approximate installed cost may be estimated for a typical diffused air system. The estimates provided herein are based on information from one diffused air equipment manufacturer(17).

The capital cost for a diffused air system involves trade-offs in the cost of blowers and the diffuser system. As a general rule, additional and/or more efficient diffusers increase the cost of the diffuser system, but decrease the size and/or number of blowers. For example, sample calculations for coarse bubble diffused air systems with various diffuser arrangements indicate the following trade-offs in diffuser system cost and blower horsepower for four different configurations capable of providing 3,500 lb O_2/day:

| | Diffuser System Cost | | Efficiency | Blower | Blower System Cost | |
Case	$/[lb O_2/day]	Total Cost	(lb/O_2/hp-hr)	(Hp)	$/[lb O_2/hp-hr]	Total Cost
1	7.73	$27,000	1.15	127	20.00	$70,000
2	7.73	27,000	1.30	112	17.60	61,600
3	8.59	30,000	1.54	95	14.86	52,000
4	11.31	39,500	1.65	88	13.83	48,400
Average	8.84		1.41		16.57	

Blower costs generally range from $250 per horsepower to $550 per horsepower for blower sizes ranging from 500 horsepower down to 100 horsepower. For the above examples, the lower blower cost of Case 4 compared to Case 1 ($21,600 difference) would more than offset the higher cost of the diffuser system of Case 4 compared to Case 1 ($12,500 difference). On average, the above examples indicate a blower cost of approximately $16 per lb O_2/day, and a diffuser system cost of approximately $9 per lb O_2/day, for a total system cost of $25 per lb O_2/day actual oxygen requirement (AOR). Note, however, that Case 4 is significantly more efficient than Case 1. On average, power costs would be 30 percent lower for Case 4 than for Case 1. At $0.05/kw-hr, the annual power cost savings would be $12,800/yr. This economic factor would result in selection of Case 4 for this application.

3.4.3.3 Mixers

Each of the anoxic zones added to the plant for denitrification must include a number of mixers to suspend the solids, as previously discussed.

The number and horsepower of mixers is directly dependent on the volume of the anoxic basin to be mixed. As described previously, the mixers are generally of the submerged propeller or submerged turbine types. The installed cost of these two types of mixers is comparable, ranging from approximately $2,300/hp for a 5 hp mixer down to $1,000/hp for a 40 hp mixer. For 20 hp mixers, an installed cost of approximately $1,300/hp may be expected. Therefore, assuming a mixing requirement of 50 hp/MG, an installed cost for mixing equipment of $65,000/MG reasonably could be expected, if 20 hp mixers are used.

3.4.3.4 Recycle Pumping

As discussed previously, biological nitrogen removal requires that a substantial recycle flow be provided from the first aerobic zone to the first anoxic zone. The sizing of this pumping system is directly related to the plant size in terms of average design influent flow.

Previous sections discuss mixed liquor recycle pumping requirements, which typically range from 1:1 to 4:1 based on plant design flow. Installed pumping system costs are generally $9 to $10 per gallon/minute(gpm) capacity, or $6,250 to $7,000 per MGD capacity for systems using pumps larger that a 2,000 gpm capacity. Systems using smaller pumps will cost significantly more, as high as $13,500 per MGD capacity for a system using 500-gpm pumps. Assuming a recycle ratio of 4:1, the recycle pumping system may be expected to cost $25,000 to $30,000 per MGD of plant capacity if relatively large capacity pumps are used.

3.4.3.5 Facility Cost Summary

The reader is again cautioned that capital costs will vary widely for biological nitrogen removal facilities. The assessments conducted in the Chesapeake Bay which were discussed above (15,16) illustrate that retrofit costs can vary from one plant to another. The unit costs described above for individual components are illustrative. They should not be used by inexperienced individuals to estimate total facility costs. Reliable estimates can be developed only by an experienced engineer utilizing site-specific information.

3.5 System Operation

3.5.1 Operational Characteristics

3.5.1.1 Nitrification

The operation and control of a suspended growth nitrification system is similar to that of a standard activated sludge system. The primary control parameters are:

o   sludge age or mean cell residence time

o   aeration basin dissolved oxygen concentration

The SRT required for nitrification is substantially higher than that for carbonaceous oxidation, particularly as wastewater temperatures drop. The sludge wasting rate must be carefully controlled to maintain the desired SRT at the current wastewater temperature. The dissolved oxygen level in the aerobic zone must be carefully monitored. Excess aeration is a waste of energy and increases operating costs. It can also result in excessive addition of oxygen to the anoxic zones by recycle pumping, thus reducing nitrogen removal. An inadequate level of aeration, on the other hand, can inhibit the nitrification process since the oxygen available will be preferentially used for carbon oxidation over nitrification.

When compared to activated sludge for secondary treatment alone, the primary operational differences nitrification creates are related to sludge production and power usage. Nitrification requires a longer SRT than activated sludge treatment for carbon oxidation due to the slower growth rate of the nitrifiers. This translates into a lower growth rate and lower associated sludge production. However, offsetting this savings is the increased aeration requirement for nitrification. The oxygen demand for nitrification is significant; approximately 4.6 pounds of oxygen are required for each pound of nitrate produced. Operation in a nitrifying mode typically increases process oxygen requirements by a factor of 50 to 100 percent over those for secondary treatment alone. This additional aeration requirement can be easily designed into a facility, but special consideration for the necessary range of operability of the aeration system must be made for plants practicing seasonal nitrification.

3.5.1.2 Denitrification

As with nitrification, denitrification has impacts on the operating characteristics of an activated sludge process. One major benefit to the activated sludge process resulting from denitrification is satisfaction of a portion of the oxygen demand for carbonaceous matter, since all or part of the readily biodegradable organic matter of the influent wastewater is consumed in the denitrification process. Generally, as much as 60 percent of the additional oxygen demand for nitrification can be recovered in the denitrification process(4).

A second major benefit to the activated sludge process is the recovery of alkalinity through denitrification. As noted previously, approximately one-half of the alkalinity consumed in the nitrification process is recovered in the denitrification process. This is of particular importance for wastewater having an alkalinity less than 200 mg/L as $CaCO_3$. A wastewater with an alkalinity of 200 mg/L as $CaCO_3$ can support the oxidation of 20 mg/L of ammonia-nitrogen to nitrate-nitrogen. However, if alkalinity is lower, the pH could drop to 6 or below. This would have an adverse effect on the nitrification process, and chemical addition would be necessary to sustain the mixed liquor pH. In many cases the alkalinity recovered through denitrification is adequate to preclude the need for pH control through chemical addition. In fact, for a low alkalinity wastewater for which nitrification only (and not nitrogen removal) is required, it may be cost-effective on a present worth basis to provide an anoxic zone rather than to add chemicals. These effects will be quantified below.

Offsetting operational factors for denitrification include the energy requirements for mixing the anoxic basin contents and for pumping the mixed liquor recycle flow. Both of these require a significant amount of energy not otherwise required for carbonaceous oxidation only. Recycle pumping can be reduced during periods of lower plant loading to control the degree of denitrification achieved and reduce recycle pumping power requirements.

3.5.1.3 Secondary Clarification

The final sedimentation, or clarification, step in a biological nitrogen removal process is operated similar to that for conventional activated sludge. Proper design and operation of the secondary clarification process is somewhat more important for a nitrogen removal process to avoid the uncontrolled loss of solids with the effluent. Not only will such loss directly impact the BOD and TSS of the plant effluent, it will also reduce the SRT in the same manner as would excessive wasting of sludge. Since the nitrification process has a minimum allowable SRT, it is possible that the nitrification process could fail if a large enough quantity of solids is lost over the weirs and/or the loss of solids occurred over a sustained period of time.

As with the typical activated sludge process, operational problems in other areas of the process can create problems in clarifier operations. The most notable is the generation of filamentous, or bulking, sludge. Although this type of sludge can produce an excellent quality plant effluent due to the filtering action of the sludge blanket, it can also overload a clarifier and create a sludge blanket that rises to the weir level. When this occurs, the effluent quality can be seriously degraded as solids pass over the weirs. One common cause of bulking sludge is an inadequate dissolved oxygen level, below 1 mg/L in the aerobic zones. Since the nitrification process substantially increases the aeration requirements, dissolved oxygen levels may drop when operating in this mode. This is particularly true when seasonal nitrification is practiced, as the seasonal conversion from carbonaceous oxidation to nitrification will increase the aeration requirements over a relatively short period of time.

Another reported cause of bulking sludge with biological nitrogen removal systems is an excessive anoxic retention period(4). If the total anoxic retention period exceeds 1.5 hours (based on total flow through the anoxic zones, including recycle flows), the development of bulking sludge is encouraged. The retention period in the first anoxic zone is controlled primarily by the mixed liquor recycle pumping rate. Therefore, within the calculated allowable recycle range for denitrification control in the first anoxic zone, the recycle rate should be maintained at the level necessary to avoid an excessive anoxic retention period. The primary anoxic basin is typically sized based on a retention period of 1.5 to 2.0 hours (based on plant influent flow only), which becomes 0.3 to 0.4 hours when a mixed liquor recycle flow of 3:1 and an RAS flow of 1:1 are included. The secondary anoxic zone is likewise sized based on a nominal retention period of 1.5 to 2.0 hours, which becomes 0.75 to 1.0 hour when an RAS flow of 1:1 is included. Therefore, a typical design would provide a total actual retention period of 1.0 to 1.4 hours, which is below the recommended limit of 1.5 hours. Of course, if plant flows are less than design (as they typically are at start up), the retention times could become excessive.

Biological nutrient removal plants are also susceptible to excessive growth of the nuisance microorganisms **Norcardia** and **Microthrix**. These organisms produce scums which will collect in great quantities on the surface of reactor vessels and clarifiers. These scums tend to be self-perpetuating and very difficult to eliminate once they develop. They not only tend to decompose (if not removed) and produce odors, but they can also overcome the scum handling facilities at the secondary clarifiers and escape with the plant effluent. The factors which favor development of problem scum are poorly understood.

Research is currently being conducted to more fully identify those factors which affect the growth of filamentous and scum-producing organisms in biological nutrient removal systems. Some experiences indicate that anoxic zones can act as "selectors" to control the growth of filamentous organisms. Selectors have been used effectively to control bulking in other activated sludge systems, and they may be effective in controlling the growth of scum-producing organisms(18). In the meantime, the design and operation of nutrient removal systems must consider that these operational problems will occur. Secondary clarifiers should be sized to take into account the possibility of bulking sludge. The biological reactor should be designed to pass biological scum to the secondary clarifier, rather than allowing it to accumulate in an uncontrolled fashion. Secondary clarifier scum removal systems should be designed to remove large quantities of biological scum from the treatment system. Scum should be directed to the solids handling train, not to the head of the plant where it can reinoculate the biological system. If these features are incorporated into the design of the facility, plant operations personnel will have the tools necessary to deal with the associated operating problems.

3.5.2 Operational Cost Considerations

3.5.2.1 Power

The single-sludge nitrogen removal process has several operational functions which will either increase or decrease the power consumption for wastewater treatment at a plant. The following paragraphs discuss each of these power-related functions and demonstrate the magnitude of power usage increase or reduction for a typical plant.

Aeration. As discussed previously, the aeration requirements of a nitrification treatment system are higher than those of a typical carbon-oxidizing activated sludge system. These additional requirements can be reduced significantly if a denitrification step (anoxic basin) is provided prior to the primary aeration basin, as in a typical single-sludge system.

Oxygen requirements for oxidation of carbonaceous organic matter and nitrogen may be estimated as follows:

$$O_c = (Q)(S_o)(8.34) \qquad\qquad (2)$$

$$O_N = (Q)(4.57)(N_o)(8.34) \qquad\qquad (3)$$

where:
O_c = oxygen required for carbonaceous matter oxidation, lb/day
Q = average daily wastewater flow rate, MGD
S_o = secondary influent BOD_5, mg/L
8.34 = conversion factor
O_N = oxygen required for nitrification, lb/day
4.57 = amount of oxygen required for nitrification, lb O_2/lb TKN
N_o = secondary influent TKN available to be nitrified, equal to the secondary influent TKN minus the TKN taken up by the activated sludge biomass, mg/L

To illustrate the impact of nitrification on aeration requirements, consider a plant with the following design characteristics:

Q = 10 MGD
S_o = 160 mg/L
N_o = 20 mg/L

Using the above equations:

O_c = (10)(160)(8.34) = 13,344 lb O_2/day = 556 lb O_2/hr
O_N = (10)(4.57)(20)(8.34) = 7,623 lb O_2/day = 318 lb O_2/hr

Assuming a typical aeration transfer efficiency of 2.0 pounds O_2 per hour per horsepower (lb/hr-hp), the following aeration horsepower requirements are indicated for this example:

Carbonaceous oxidation = 278 hp
Nitrification = 159 hp

For denitrification in the anoxic zones, approximately 2.86 pounds of oxygen demand are satisfied per pound of nitrate consumed. If complete denitrification is assumed, the following oxygen demand will be satisfied in the anoxic zones:

O_2 Demand Satisfied = (2.86)(20 mg/L)(10 MGD)(8.34) = 4,770 lb/day (or 199 lb/hr)

At an aerator efficiency of 2.0 lb O_2/hr-hp, this is equivalent to a reduction in aerator horsepower of 99 hp.

Assuming a motor efficiency of 0.9, this translates into the following annual electrical power reduction:

Power reduction = (99 hp)(0.746 kw/hp)(24)(365)/(0.9) = 720,000 kwh/yr

At a power cost of $0.07 per kwh, this equals $50,400 per year in power savings over a system that includes only nitrification.

<u>Mixing</u>. Partially offsetting the aeration power savings determined above is the energy required for mixing the anoxic zones. Using typical design nominal retention periods of 2.0 hours for each of the anoxic zones, the volume of each zone is determined as follows:

Reactor size $\quad$ = (10 MGD) (2.0 hr) / (24 hr/day) = 0.83 MG

Assuming that the required mixing energy is 50 hp/MG, the mixing energy requirements are:

Primary reactor mixing $\quad$ = (50 hp/MG) (0.83 MG) $\quad$ = 42 hp
Secondary reactor mixing $\quad$ = (50 hp/MG) (0.83 MG) $\quad$ = 42 hp
Total mixing energy $\qquad\qquad\qquad\qquad\qquad\qquad$ = 84 hp

Assuming a power cost of $0.07 per kwh, this translates into the following annual electrical power cost:

Power $\quad$ = (84 hp) (0.746 kw/hp) (24) (365) = 550,000 kwh/yr
Cost $\quad$ = (550,000 kwh/yr) ($0.07/kwh) = $38,400/yr

<u>Recycle Pumping</u>. An additional offsetting cost for operating a single-sludge system is the pumping of recycled mixed liquor to the first anoxic zone. This pumping system requires a high capacity, but the pumping head is low due to the small difference in water level between the basins and the short piping length involved. Continuing with the above example, if it is assumed that the mixed liquor recycle ratio is 3:1 (pumping capacity is therefore 30 MGD), the pumping head is 10 feet, and the pump efficiency is 0.75, the following power requirement is determined:

$$\text{Pump brake hp} \quad = \frac{(30 \text{ MGD}) \times (694 \text{ gpm/MGD}) \times (10 \text{ ft})}{(3,960)(0.75)} = 70 \text{ hp}$$

where 3,690 is a dimensional conversion factor for the units used in this calculation(19).

Assuming a motor efficiency of 0.9 and a power cost of $0.07 per kwh, this translates into the following annual power cost:

Power $\quad$ = (70 hp) (0.746 kw/hp) (24) (365)/(0.9) = 508,275 kwh/yr
Cost $\quad$ = (508,275 kwh/yr) ($0.07/kwh) = $35,600/yr

From the above calculations, the additional operating cost for recycle pumping and mixing of $74,000 for this example is nearly offset by a cost reduction for aeration of $50,400 per year for the reduction in BOD loading resulting from the denitrification process. Other results will be obtained in other situations. However, as a general rule, the single-sludge system does not increase operating costs appreciably over those for nitrification alone.

3.5.2.2 Alkalinity

Another operational cost consideration for the single-sludge process is that related to alkalinity and pH control. As previously discussed, the denitrification process reclaims approximately one-half of the alkalinity consumed in the nitrification process. For a wastewater that is relatively high in alkalinity, this is of no consequence. However, for wastewaters with alkalinities below 200 mg/L as $CaCO_3$, alkalinity return by denitrification may significantly reduce or eliminate chemical addition for pH control.

The two most common chemicals used for pH control of mixed liquor are sodium hydroxide (NaOH) and calcium hydroxide, or lime (CaO). Sodium hydroxide is commonly purchased as a 50-percent solution, which is fed to the mixed liquor at a controlled rate using chemical metering pumps. Lime, on the other hand, is purchased in a dry form either as quicklime (CaO) or hydrated lime ($Ca[OH]_2$). Quicklime must be "slaked" prior to addition to the mixed liquor, which requires a special piece of equipment referred to as a "slaker." The slaking process mixes the quicklime with water in a controlled process to produce hydrated lime. This approach is usually found to be cheaper than purchasing the lime in a slaked form.

Continuing the previous example, the chemical cost for pH control that would be eliminated by the denitrification process may be estimated. Assume again a plant flow of 10 MGD and an influent nitrogen level of 20 mg/L NH_4^+-N. If essentially complete nitrification occurs in the aerobic zone, the following amount of alkalinity will be consumed:

$$\text{Consumed Alkalinity, as } CaCO_3 \quad = (7.2 \text{ lb } CaCO_3/\text{lb } NO_3\text{-N}) \text{ (20 mg/L) (10 MGD) (8.34)}$$
$$\text{(Nitrification)} \quad = 12,010 \text{ lb/day}$$

If complete denitrification occurs in the anoxic zone, the following amount of alkalinity will be produced:

$$\text{Produced Alkalinity, as } CaCO_3 \quad = (3.6 \text{ lb } CaCO_3/\text{lb } NO_3\text{-N}) \text{ (20 mg/L) (10 MGD) (8.34)}$$
$$\text{(Denitrification)} \quad = 6,005 \text{ lb/day}$$

The alkalinity consumed in nitrification may or may not require replacement through chemical addition, depending on the beginning alkalinity of the wastewater. However, if chemical replacement of alkalinity is needed, the alkalinity produced by denitrification represents a savings in chemical costs. For example, if sodium hydroxide is used for pH control, the following calculation indicates the quantity of sodium hydroxide that would be saved ($CaCO_3$ equivalent weight = 50; NaOH equivalent weight = 40):

$$\text{NaOH saved} \quad = (40) \text{ (6,005 lb/day) / (50)} = 4,804 \text{ lb/day}$$

The bulk cost of sodium hydroxide generally ranges from $250 to $350 per dry ton. At a cost of $300 per dry ton, the annual savings in chemical cost would be:

$$\text{Cost Savings} \quad = (\$300/\text{dry ton}) \text{ (4,804 lb/day) (365)/(2,000 lb/ton)} = \$263,000/\text{yr}$$

When compared to the net energy cost of approximately $24,000 per year as calculated previously for this example, it is evident that the use of denitrification to minimize the chemical costs for a low alkalinity wastewater can be a significant economic factor.

3.5.2.3 Sludge Disposal

In general, the sludge disposal costs for a single sludge biological nitrogen removal system will be equal to or less than those for secondary treatment alone. No extra chemicals which result in additional quantities of sludge for disposal are added to the process. In fact, since relatively long SRTs must be maintained to ensure stable nitrification, the quantity of sludge produced by the process may be somewhat reduced. This results from the increased endogenous respiration (cell death and decay) which will occur in long SRT systems. Thus, the assumption of no impact on sludge disposal costs is conservative when evaluating a single-sludge biological nitrogen removal system.

3.6 Full-Scale Experience

3.6.1 General

In the 1970s, biological nitrogen removal plants of the separate-stage type using methanol as a carbon source were considered state-of-the-art. However, as experience and understanding increased the single-sludge systems increased in popularity. These systems have also been found to be easier to operate, being essentially a modification of the popular activated sludge secondary treatment system. Another reason for their popularity is the ease of incorporating biological phosphorus removal into the treatment system, as discussed in Chapter 7. This is significant because effluent permits which require nitrogen removal often require phosphorus removal as well.

Although widely considered in the 1970's, relatively few methanol-driven separate stage denitrification systems have been constructed and remain in operation. Most operating systems are of the downflow, deep bed, granular filter type illustrated in Figure 3-2. However, full-scale examples of the fluidized bed(20) and suspended growth systems also exist. Since the late 1970's, numerous single-sludge carbon oxidation- nitrification-denitrification facilities were constructed and are currently in operation. Some single-sludge systems exist which are followed by downflow, deep bed denitrifying granular filters used for polishing. This section presents case histories of both separate stage and single-sludge biological nitrogen removal facilities.

3.6.2 Case Studies

In this section actual operating biological nitrogen removal plants and pilot studies are described, along with operating performance data for those plants and studies. These plants include many of the different process configurations discussed in this chapter, as follows:

Type	Plant
Separate stage, packed bed	Hooker Point WWTP, Tampa, Florida
Separate stage, fluidized bed	Reno-Sparks WWTF, Nevada
Separate stage, suspended growth	River Oaks AWTP, Hillsborough County, Florida
Single-sludge, primary anoxic/aerobic zone	Largo WWTP, Largo, Florida Fayetteville WWTP, Fayetteville, Arkansas Virginia Initiative Plant(VIP), Hampton Roads,Virginia Landis Sewage Authority, Vineland, New Jersey
Single-sludge, primary and secondary anoxic/aerobic zones	Palmetto WWTP, Palmetto, Florida Eastern Service Area WWTP, Orlando, Florida
Sequencing Batch Reactor	Del City WWTP, Del City, Oklahoma

Several of the plants discussed in the following paragraphs also practice phosphorus removal in addition to nitrogen removal. The discussions in this chapter will concentrate on the nitrogen removal aspects of those plants. The phosphorus removal features and performance of some of these plants will be discussed in detail in Chapter 7.

3.6.2.1 Hookers Point Wastewater Treatment Plant, Tampa, Florida

Facility Description. The Hookers Point Wastewater Treatment Plant (WWTP) is a 60-MGD facility (to be expanded to 96-MGD) including preliminary treatment, primary treatment, biological treatment, post aeration, and effluent disinfection. A schematic of the biological treatment portion of the plant is presented in Figure 3-10. The secondary treatment system includes carbonaceous oxidation/nitrification using high purity oxygen (HPO), and a separate-stage packed bed denitrification system using methanol as a carbon source. The HPO system flow scheme includes an initial set of reactors for BOD_5 removal, intermediate clarifiers, a second set of reactors for nitrification, and final clarifiers. Therefore, the Hookers Point WWTP is a "three-stage" system, as described earlier in this chapter.

The maximum recycle ratio to the first aerobic zone is 0.78:1 (with respect to the plant influent). The maximum recycle ratio to the second aerobic zone (nitrification) is 1:1. The design loading on the first aerobic zone is 1.2 lbs BOD_5/lb MLVSS under aeration/day, with a design MLVSS of 3,900 mg/L. The second aerobic zone has a design MLVSS of 2,500 mg/L and a hydraulic retention time with respect to influent wastewater flow of 2 hours. The separate-stage denitrification filters consists of 6 feet of coarse sand, loaded at an average rate of 1 gpm/ft^2. The empty bed detention time is 45 minutes at average flow.

Effluent Limits. Effluent limits for the Hookers Point WWTP are 5 mg/L for both $TBOD_5$ and TSS. The total nitrogen (TN) limit is 3 mg/L on an annual average basis and the total phosphorus (TP) limit is 7.5 mg/L.

Wastewater Characteristics. The Hookers Point WWTP receives municipal wastewater from the entire municipality of Tampa, as well as a significant industrial component from sources such as breweries. The wastewater characteristics of interest are:

Parameter	Average
$TBOD_5$, mg/L	224
TSS, mg/L	221
NH_3-N, mg/L	32

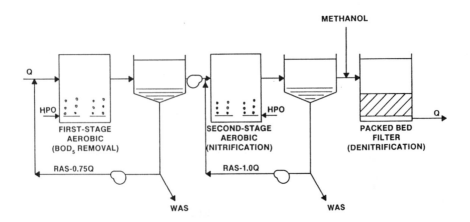

Figure 3-10. Biological treatment flow scheme for Hookers Point WWTP, Tampa, Florida.

<u>Operating Results</u>. The plant has been loaded to its design values and has demonstrated excellent performance. Plant effluent has generally been in full compliance with its discharge standards, as listed above. The separate stage denitrification system has performed very well, achieving an average effluent total nitrogen concentration of 2.8 mg/L, which is below the discharge standard of 3 mg N/L. A probability plot of monthly average effluent total nitrogen concentrations from this plant is presented in Figure 3-18.

<u>Summary</u>. The excellent performance of the Hookers Point plant demonstrates the capability of a separate stage packed bed denitrification system to consistently achieve low effluent TN concentrations (less than 3 mg/L). In addition, the process is capable of producing an effluent low in $TBOD_5$ and TSS through the filtering action of the packed bed denitrification system.

3.6.2.2 Reno-Sparks Wastewater Treatment Facility, Cities of Reno and Sparks, Nevada

<u>Facility Description</u>. Figure 3-11 depicts the flow schematic for the Reno-Sparks Wastewater Treatment Facility (WWTF). The liquid process train consists of preliminary treatment, primary treatment, phosphorus and BOD removal in a Phostrip^{TM} system, nitrification in nitrifying trickling filters, denitrification in methanol-driven upflow fluidized bed reactors, post aeration, effluent filtration, and disinfection. The solids handling system consists of thickening, anaerobic digestion, and dewatering. The plant has been expanded and upgraded step-wise over the years, with the nitrogen removal facilities becoming available in 1989.

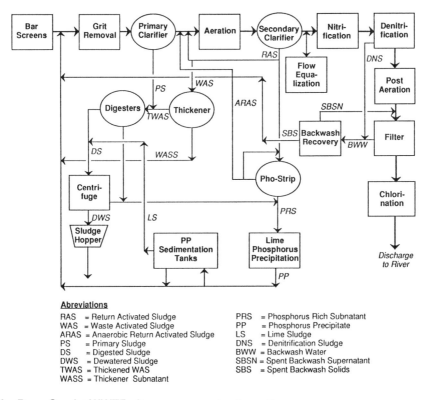

Figure 3-11. Reno-Sparks WWTF nitrogen removal schematic.

73

Effluent Limits. The Reno-Sparks WWTF discharge limits, based on a monthly average, are as follows:

Parameter	Discharge Limit
Flow, MGD	40
BOD_5 (inhibited), mg/L	10
BOD_5 (uninhibited), mg/L	20
Suspended Solids, mg/L	20
Total N, mg/L	5
Total P, mg/L	0.4[a]

[a]Based on flow of 40 MGD; mass limitation is 134 pounds per day.

Wastewater Characteristics. The average influent characteristics for 1986 for the plant are listed below. The actual values experienced are still somewhat less than the design values.

Parameter	Actual	Design
Flow (max. month), MGD	26.5	40
BOD_5 (inhibited), mg/L	156	--
BOD_5 (uninhibited), mg/L	188	275
Suspended Solids, mg/L	177	250
Total N, mg/L	--	30
Total P, mg/L	8.5	10

Operating Results. The final effluent characteristics for 1986 are presented in the following table. Also included in the table are more current data (July 1989-July 1990) for total nitrogen and phosphorus. During the July 1989-July 1990 period, the monthly average for total nitrogen never exceeded the 5 mg/L limit; the highest monthly measurement was only 4.6 mg/L. Further, the measurements for May, June, and July 1990 were 1.38, 1.43, and 1.11 mg/L, respectively. These new levels should be more representative of the capability of the process, since it is believed that the reactors were not operating properly until March 1990.

Parameter	1986 Average	July '89-July '90 Average
Flow, MGD	--	--
BOD_5 (inhibited), mg/L	5.5	--
BOD_5 (uninhibited), mg/L	10	--
Suspended Solids, mg/L	7.3	--
Total N, mg/L	--	2.45
Total P, mg/L	0.33	0.21

A probability plot for effluent total nitrogen concentrations for July 1989 through July 1990 is presented in Figure 3-18.

Summary. The Reno-Sparks WWTF has been producing effluent well within its permitted discharge limits. While the reactors were not originally operating properly, they still managed to produce effluent with acceptable total nitrogen levels. Currently, the nitrogen removal system is producing excellent results, with effluent total nitrogen concentrations as low as 1.11 mg/L.

3.6.2.3 River Oaks Advanced Wastewater Treatment Plant, Hillsborough County, Florida

Facility Design. The River Oaks Advanced Wastewater Treatment Plant (AWTP) was upgraded in three phases beginning in 1986 to increase its capacity from 3 MGD to 10 MGD. Phase one of the upgrade included a denitrification system utilizing aerobic stabilization. Methanol was used as the carbon source for this system. The upgraded plant also includes headworks, primary clarification, aeration, secondary clarification, and final flocculation/clarification (after denitrification).

Figure 3-12 depicts the denitrification process. A mixing area is provided at the head of the denitrification system to allow for the addition of methanol and to provide a point of introduction for the return denitrified sludge. The two 0.065 MG denitrification tanks, operated in parallel, consist of 16 cells each. These cells are divided into two zones, anoxic and aerobic. The first 10 cells are designed to operate only in the anoxic mode, the next four can be operated in either the anoxic or aerobic mode, and the last two can be operated only in the aerobic mode. The design criteria called for an HRT of 3.8 hours; a SRT of 30 days; a F/M (NO_x-N) of 0.12 kg NO_x-N/kg MLVSS-day; and an influent NO_x-N concentration of 25 mg/L.

Effluent Limits. Since the plant's discharge enters Tampa Bay, its effluent limits are very stringent -- the most stringent for any plant in Florida. The discharge limits are based on pounds of mass per day as follows:

Parameter	Mass Load (lbs/day)
BOD_5	197
TSS	197
Total N	117
Total P	39

The design maximum month flow of 12 MGD corresponds to a total nitrogen limit of 1.2 mg/L.

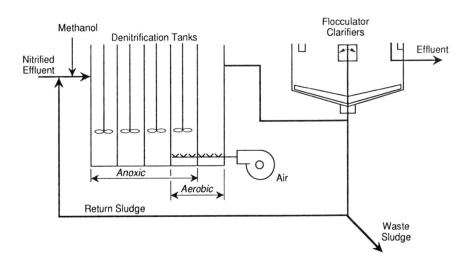

Figure 3-12. River Oaks AWTP separate stage suspended growth process.

75

<u>Wastewater Characteristics</u>. The plant loadings experienced from August 1988 to July 1989 were, in general, somewhat lower than anticipated in the design. The design and observed values for the parameters of interest can be seen in the following table. These values are based on the peak month conditions.

Parameter	Design	Actual
Flow, MGD	10	9.5
BOD_5, mg/L	200	176
TSS, mg/L	275	149
TKN, mg/L	31	37.6
Total P, mg/L	9	7.0

<u>Operating Results</u>. The plant performance during the August 1988 to July 1990 period was excellent. The plant consistently met its discharge limits, with the exception of one week when the total nitrogen concentrations averaged 3.2 mg/L. This was an anomaly, since the plant ran out of methanol during that time period. The table below summarizes the operating results for the August 1988 to July 1990 period.

Parameter	Average Effluent	Maximum Monthly Effluent
Flow, MGD	7.7	9.5
BOD_5, mg/L	<2	2
TSS, mg/L	<2	2
Total N, mg/L	1.0	1.4
Total P, mg/L	0.24	0.40

During this period the SRT averaged 16 days and ranged from 8 to 25 days. The MLTSS averaged 4,350 mg/L and ranged from 2,750 to 5,650 mg/L. The denitrified effluent turbidity averaged 0.9 NTU and ranged from 0.5 to 1.2 NTU. A probability plot for effluent total nitrogen concentrations from this plant is presented in Figure 3-18.

<u>Summary</u>. The separate stage, suspended growth denitrification system at the River Oaks AWTP has demonstrated excellent results in removing nitrogen to concentrations below 1 mg/L. The plant is operating as anticipated in the design and continues to meet its stringent discharge limits.

3.6.2.4 Largo WWTP, Largo, Florida

<u>Facility Description</u>. The Largo WWTP consists of three parallel treatment trains providing a total plant design capacity (average flow) of 15 MGD. The plant includes preliminary treatment, primary treatment, secondary treatment, effluent filtration, and disinfection. The plant uses the A^2/O process to remove both nitrogen and phosphorus, as illustrated in Figure 3-13 (see Chapter 7 for a detailed discussion of this process). The process is similar to the single-sludge Bardenpho process, described earlier in this chapter, in that it includes an aerobic zone with mixed liquor recycle to a preceding anoxic zone. However, it uses only the first anoxic and aerobic zones. Unlike the Bardenpho process, the A^2/O process is a high-rate process, typically operated at an SRT of less than 10 days. In addition, the mixed liquor recycle ratio is only 1:1 to 2:1 with respect to plant influent flow rate. Hydraulic retention times (HRT) of 0.8 hour in the anaerobic zone, 0.5 hour in the anoxic zone, and 2.9 hours in the aerobic zone (total HRT of 4.2 hours) are provided at design flow. The wastewater temperature is typically higher than 20°C, which allows nitrification to proceed even at the relatively low plant SRTs.

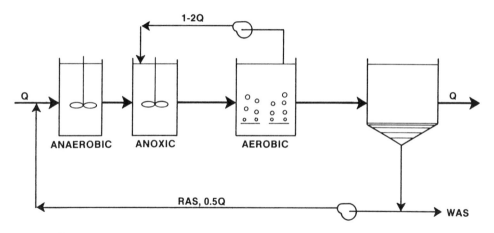

Figure 3-13. A^2/O Process as used in Largo, Florida.

Effluent Limits. TBOD$_5$ and TSS effluent limitations are each 5 mg/L. Nitrogen limitations are established for the following three frequencies:

	Total Nitrogen (mg/L)
Annual average	8
Monthly average	12
Weekly average	18

The plant is also restricted on effluent ammonia-nitrogen to 2 mg/L and 3 mg/L for monthly and weekly averages, respectively.

Wastewater Characteristics. The Largo plant receives a typical, medium strength municipal wastewater. The influent wastewater characteristics for the plant are as follows:

Parameter	Average	Range
TBOD$_5$, mg/L	200	113 - 375
TSS, mg/L	325	143 - 511
TKN (maximum), mg/L	30	---
NH$_3$-N (maximum), mg/L	20	---

Operating Results. The average plant flow over the period of January 1984 to November 1987 was 9.9 MGD, which is approximately two-thirds of the plant design capacity. The MLSS concentration was generally held below 3,000 mg/L. On average the plant has performed within the permit limitations for TBOD$_5$ and TSS. The TBOD$_5$ in the plant effluent has averaged 5 mg/L and the TSS has averaged 4 mg/L.

Due to the high rate nature of the A^2/O process used at Largo, the removals of nitrogen are not expected to be as high as with the Bardenpho-type process. However, the Largo plant averaged a monthly effluent total nitrogen level of 7.7 mg/L, meeting the monthly average effluent standard for 43 of 44 months during the subject period. A probability plot of effluent total nitrogen concentrations from this plant is presented in Figure 3-18.

Summary. The Largo plant is an example of the use of the high-rate A^2/O process to provide partial nitrogen removal. A key factor in the successful removal of nitrogen in a high rate system at this plant is the high year-round wastewater temperature. This particular system would not perform as well with respect to nitrogen removal if the minimum wastewater temperature was more typical of a plant in a cooler climate. However, its performance indicates what is possible with a plant of this configuration.

3.6.2.5 Fayetteville WWTP, Fayetteville, Arkansas

Facility Description. The Fayetteville WWTP is a 17-MGD facility, including preliminary treatment, primary treatment, secondary treatment, effluent disinfection, and post-aeration. A schematic of the biological reactor may be found in Figure 3-14.

The plant is designed around the A/O process, which is discussed in detail in Chapter 7. Normally, the A/O process includes an anaerobic zone followed by an aerobic zone, and it is intended primarily for phosphorus removal. However, at the Fayetteville plant the conversion of ammonia to nitrate is also required, and so the plant is designed to nitrify. When this is done with an A/O plant, it is often desirable to also add a partial denitrification step to remove a portion of the nitrogen. This in turn reduces the nitrate loading on the anaerobic zone due to the RAS, thereby enhancing phosphorus removal. This modification of the A/O process is the A^2/O process, as discussed above. At the Fayetteville plant, flexibility has been provided to allow operation in either the A/O mode or the A^2/O mode.

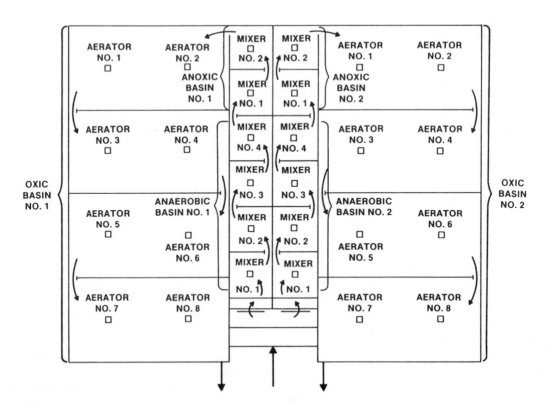

Figure 3-14. Fayetteville WWTP: aeration basins flow pattern.

Although performance data on the full-scale facility is not yet available, a 1-gpm pilot-scale plant was operated during 1985. The pilot scale facility was carefully designed to accurately simulate the full-scale plant and has operated on existing Fayetteville primary effluent. The performance results discussed in the following are therefore based on the pilot study(21).

Effluent Limits. The effluent limits (monthly averages) are seasonal in nature, as follows (expressed as mg/L):

Period	$TBOD_5$	TSS	NH_3-N	DO	Total Phosphorus
April to November	5	5	2	7.8	1
December to March	10	10	5	10.2	1

Wastewater Characteristics. The Fayetteville plant receives a relatively high-strength wastewater, including both municipal and industrial components. The characteristics of the primary effluent used in the pilot study were as follows:

Parameter	Average	Range
$TBOD_5$, mg/L	139	115 - 181
TSS, mg/L	93	81 - 122
NH_3-N, mg/L	11.2	5.9 - 15.5
Temperature	--	10°C - 25°C

Operating Results. The pilot plant was operated at SRTs ranging from 2.8 to 13.7 days, with the longer SRTs occurring in the winter. The plant total HRT was maintained at 6 to 8 hours for design average loading conditions. The combined HRT in the anaerobic and anoxic zones was between 1 and 2 hours. The MLSS concentration ranged from 1,370 to 3,100 mg/L. The pilot plant was generally operated in the A/O mode, although the A^2/O mode was used occasionally.

The pilot plant operated well with respect to $TBOD_5$ and TSS removal. The effluent concentrations were generally below 5 mg/L. The effluent nitrogen levels varied from 3.7 to 15.1 mg/L, with the lower levels resulting from operation in the A^2/O mode (denitrification). A probability plot of effluent total nitrogen concentrations from this plant is presented in Figure 3-18.

Summary. The Fayetteville pilot plant, although intended for phosphorous removal and nitrification only, demonstrates the capability to reduce effluent nitrogen concentrations to relatively low levels when operating in the A^2/O mode.

3.6.2.6 Virginia Initiative Plant (VIP) Pilot Study, Hampton Roads Sanitation District (HRSD), Virginia

Facility Description. The Lamberts Point WWTP is being expanded and upgraded by the Hampton Roads Sanitation District (HRSD) to provide 40-MGD of secondary treatment capacity. The new plant includes influent pumping, preliminary treatment, primary treatment, secondary treatment, and effluent disinfection. The secondary treatment system includes nitrogen and phosphorus removal capabilities. The design criteria for the full-scale plant were developed through an extensive pilot plant study(22). The nutrient removal process resulting from the pilot study is called the Virginia Initiative Plant (VIP) process. The new plant has been likewise named the VIP. The VIP process is shown schematically in Figure 3-15.

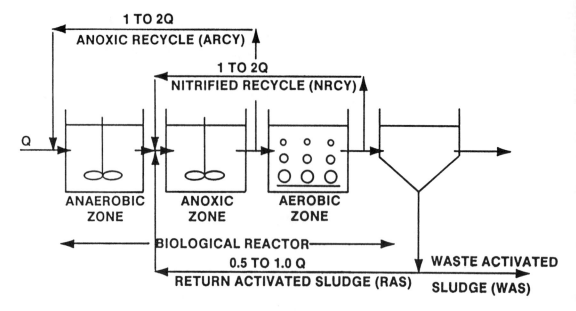

NOTE: A STAGED REACTOR CONFIGURATION IS
PROVIDED BY USING AT LEAST TWO COMPLETE
MIX CELLS IN SERIES FOR EACH ZONE OF THE
BIOLOGICAL REACTOR.

Figure 3-15. VIP Process.

Like other biological nutrient removal systems, the VIP process includes three zones: anaerobic, anoxic, and aerobic. However, in the VIP process the RAS is recycled to the anoxic basin (downstream of the anaerobic basin) and a denitrified mixed liquor flow stream is recycled at a rate of 1:1 to 2:1 from the anoxic zone back to the anaerobic zone. This process is intended to improve operation of the anaerobic zone by reducing the potential of nitrate loading on the anaerobic zone by recycling anoxic effluent, rather than RAS, to that zone. The full-scale VIP has been designed to remove phosphorus year-round and nitrogen on a seasonal basis (during summer). It is currently under construction. The anaerobic and anoxic zones will constitute 34 percent of the total reactor volume, with an HRT in the overall process of 6.5 hours at the design average flow of 40 MGD.

Effluent Limits. The plant effluent permit for the VIP requires a $TBOD_5$ and TSS of not more than 30 mg/L each, but it does not limit either nitrogen or phosphorus in the effluent. As a result, the funding for the plant was limited to that typical for a secondary treatment facility only (i.e., no nutrient removal). However, due to concerns over the water quality of Chesapeake Bay, the HRSD chose to incorporate partial nutrient removal to the extent possible within the limits of conventional reactor sizing for secondary treatment.

The following goals were established for nutrient removal, above and beyond the effluent permit limitations:

o   Phosphorous   67 percent removal, year-round

o   Nitrogen   70 percent removal for wastewater temperatures above 20°C, less for
    lower temperatures.

Wastewater Characteristics. The wastewater treated in the pilot study and by the VIP plant is a relatively weak domestic wastewater. The wastewater characteristics of interest are as follows:

Parameter	Average	Range
TBOD$_5$, mg/L	142	109 - 199
TSS, mg/L	133	98 - 152
Total Nitrogen, mg/L	25.0	21.2 - 29.3
Temperature	--	13°C - 25°C

Operating Results. During the pilot study, the plant HRT ranged from 4 to 8 hours, and the SRT ranged either from 5 to 6 days or from 10 to 11 days, depending on wastewater temperature. The MLSS concentration ranged from 1,200 to 3,000 mg/L. Nitrification was generally maintained throughout the study with two exceptions, one related to process upset and the other to an intentional reduction of SRT. The plant effluent TBOD$_5$ and TSS concentrations were well below that required by the effluent permit. Average values were 8 mg/L for TBOD$_5$ and 10 mg/L for TSS. This excellent performance is partially attributable to an oversized secondary clarifier in the pilot plant. The effluent total nitrogen level was consistently below 10 mg/L during the periods when full nitrification was maintained (most of the study duration). A probability plot of effluent total nitrogen concentrations from this plant is presented in Figure 3-18.

Summary. The primary emphasis of the VIP pilot study was on the removal of phosphorous, which is addressed in Chapter 7 of this manual. However, the pilot study demonstrated the capability of a high-rate biological nutrient removal facility, comparable in size and cost to a secondary treatment facility, to achieve significant removals of nitrogen.

3.6.2.7 Landis Sewerage Authority WWTP, Vineland, New Jersey

Facility Description. The Landis Sewerage Authority WWTP is an 8.2 MGD (annual average) facility that includes preliminary treatment, primary treatment, secondary treatment, effluent disinfection, and an effluent land application system. A schematic of the secondary treatment portion of the plant is presented in Figure 3-16. The secondary treatment system includes denitrification basins followed by mechanically-aerated carbonaceous oxidation/nitrification basins. Nitrified mixed liquor is recycled to the head end of the denitrification basins, along with return activated sludge.

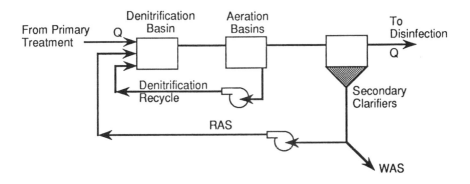

Figure 3-16. Landis Sewerage Authority WWTP denitrification schematic.

81

Because of the high BOD_5 concentration in the primary effluent (approximately 260 mg/L), no additional carbon source is needed for denitrification. The average recycle ratio for the denitrification recycle is 3.0 (with respect to the plant influent). The design loading onto the secondary system is 26,200 lb/d BOD_5, 1,700 lb/d NH_3-N, and 3,000 lb/d TKN. Other design conditions include: total MLSS concentration of 3,250 mg/L, SRT (aeration) of 11.4 days, and a low temperature of 10°C.

Effluent Limits. The monthly average effluent limits for the Landis Sewerage Authority WWTP are:

Parameter	Average
BOD_5, mg/L	30
TSS, mg/L	30
NH_3-N, mg/L	0.5
NO_3-N, mg/L	10

Wastewater Characteristics. The WWTP receives municipal wastewater from the surrounding towns, as well as a significant food processing load. The wastewater characteristics of interest are:

Parameter	Average
$TBOD_5$, mg/L	399
TSS, mg/L	223
NH_3-N, mg/L	20
TKN, mg/L	40

Operating Results. During a May 1990 test, the plant demonstrated excellent performance when loaded to near its design values. For this test, the final effluent values averaged 2.7 mg/L BOD_5, 4.1 mg/L TSS, 0.2 mg/L NH_3-N, and 4.4 mg/L NO_3-N. The effluent parameters represented in the following table are those for normal operating conditions.

Parameter	Average
$TBOD_5$, mg/L	6.6
TSS, mg/L	4.8
NH_3-N, mg/L	0.3
TKN, mg/L	3.4

Summary. The excellent performance of the Landis Sewerage Authority WWTP demonstrates the capability and reliability of the single-sludge, single anoxic zone process. Even with the plant loaded near design capacity the plant was able to maintain effluent levels well within its discharge permit.

3.6.2.8 Palmetto WWTP, Palmetto, Florida

Facility Description. The Palmetto WWTP is an advanced treatment facility with an average capacity of 1.4 MGD. The plant includes preliminary treatment, primary clarification, secondary treatment, effluent filtration and effluent disinfection. The secondary treatment system uses a Bardenpho process for nutrient removal. In a minor modification of this process, the primary sludge at Palmetto is discharged to the RAS wet well to become incorporated into the mixed liquor. In turn, a portion of the RAS is diverted to the primary clarifiers, which serve as a portion of the anaerobic zone for the Bardenpho process. A process schematic for the facility is presented in Figure 3-17.

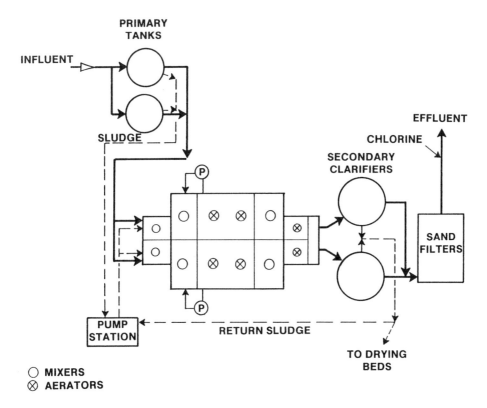

Figure 3-17. Palmetto WWTP liquid process train.

The Palmetto Bardenpho system is designed for a total HRT at design flow of 11.6 hours. This HRT is divided among the various zones as follows:

o   Anaerobic zone   1.0 hr
o   First anoxic zone   2.7 hr
o   First aerobic zone   4.7 hr
o   Second anoxic zone   2.2 hr
o   Second aerobic zone   1.0 hr

The mixed liquor recycle pumping system has the capability to return up to 400 percent of the plant flow from the first aerobic zone to the first anoxic zone.

Effluent Limits.   The effluent monthly average permit limits (mg/L) for the Palmetto plant are as follows:

Parameter	Limit
TBOD$_5$	5
TSS	5
Total Nitrogen	3
Total Phosphorus	1

Wastewater Characteristics. The Palmetto wastewater is a primarily domestic wastewater of medium strength. The parameters for which the plant was designed are somewhat lower than the values observed during the period of January 1984 through November 1987. The design and observed values for the parameters of interest are as follows:

Parameter	Design Value	Observed Value Average	Range
$TBOD_5$, mg/L	270	158	87 - 232
TSS, mg/L	250	135	70 - 224
TKN, mg/L	43	33.1	15.1 - 45.9
Temperature	---	---	18°C - 25°C

Operating Results. During the period of operation from January 1984 to November 1987 the plant was loaded at and above its design hydraulic capacity, but was under loaded with respect to organic and nutrient loadings. The average daily plant flow ranged from 0.74 to 2.44 MGD, versus a design capacity of 1.4 MGD, but the plant $TBOD_5$ loading was only 54 percent of design.

In order to meet the stringent effluent nitrogen limit, the plant was operated at an SRT ranging from 14 days (in summer) to 20 days (in winter). The average MLSS concentration was 4,090 mg/L, which is higher than the design value of 3,500 mg/L. The Palmetto plant routinely met its effluent permit limitations during the subject period. The relatively long SRT, along with wastewater temperatures generally in excess of 20°C, have allowed this plant to achieve excellent removals of nitrogen. A probability plot of effluent total nitrogen concentrations from this plant is presented in Figure 3-18.

Summary. The Palmetto plant is an excellent example of a successfully operating Bardenpho nutrient removal plant. The excellent operating results emphasize the need to provide an adequate SRT capability in the plant design if extensive nitrogen removal is needed.

3.6.2.9 Eastern Service Area WWTP, Orlando, Florida

Facility Description. The Eastern Service Area WWTP has an average design capacity of 6 MGD. The plant includes preliminary treatment, secondary treatment, effluent filtration, and effluent disinfection. This plant uses a Bardenpho process for biological nutrient removal, as described in previous sections of this chapter. The plant is designed to operate with an MLSS of 4,500 mg/L.

The Bardenpho process at the Eastern Service Area WWTP is designed with a total HRT of 15.0 hours under peak month flow conditions (7.8 MGD). This HRT is provided by the various zones in the treatment train as follows:

- o Anaerobic zone   2.0 hr
- o First anoxic zone   2.8 hr
- o First aerobic zone   8.5 hr
- o Second anoxic zone   1.5 hr
- o Second aerobic zone   0.2 hr

Effluent Limits. The Eastern Services Area WWTP has several options for effluent discharge and, therefore, has varying permit limits. The most restrictive effluent limitations are the interim limits as follows (in mg/L):

Frequency	TBOD$_5$	TSS	Total Nitrogen
Annual average	5	5	3
Monthly average	8	8	5

Wastewater Characteristics. The wastewater treated at the Eastern Service Area WWTP is primarily domestic in origin and of medium strength. The design parameters match fairly closely the values observed during 1985 and 1986, as follows:

Parameter	Design Value	Observed Value Average	Range
TBOD$_5$, mg/L	190	167	105 - 241
TSS, mg/L	191	149	96 - 224
TKN, mg/L	35	30.7	22.1 - 40.1

The wastewater temperature ranges are not available, but should be similar to those for the Palmetto plant (18°C to 25°C), since both plants are located in central Florida.

Operating Results. The Eastern Service Area WWTP has experienced flows ranging from 2.5 to 4.4 MGD, substantially less than the plant design capacity of 6 MGD. This, coupled with the near-design concentrations of influent loading constituents, indicates that the plant has been under loaded with respect to design capacity. The plant performance has been excellent, meeting all of the plant effluent limits. The TBOD$_5$ and TSS concentrations have been consistently below 5 mg/L each, and the effluent total nitrogen concentration has been below 3 mg/L. A probability plot of effluent total nitrogen concentrations from this plant is presented in Figure 3-18.

Summary. The Eastern Service Area plant confirms the capability of the Bardenpho process to achieve a high degree of nitrogen removal, if properly designed and operated. As discussed previously, a key factor in this success lies in the sizing and design of the aerobic cell and plant SRT to ensure that conditions are met for complete nitrification.

3.6.2.10 Del City WWTP, Del City, Oklahoma

Facility Description. The Del City WWTP is a 3.0 MGD sequencing batch reactor (SBR) facility which is operated in a nitrogen removal mode for energy conservation purposes and to control sludge settleability. The system consists of preliminary treatment (comminutor and grit removal), the SBR, ultraviolet disinfection, and aerobic sludge digestion. The SBR system consists of two circular basins, each 113 ft. in diameter with a 22 ft. total water depth. The total hydraulic residence time is 26 hours at the design average flow. Mixing and oxygen transfer are provided by jet aerators. Effluent discharge is provided by a floating, submerged, effluent decanter.

Effluent Limits. The Del City WWTP was designed to provide secondary treatment to achieve effluent concentrations of 30 mg/L TBOD$_5$ and 30 mg/L TSS.

<u>Wastewater Characteristics</u>. Design and actual wastewater characteristics are summarized as follows:

Parameter	Design	Actual
Flow, MGD	3.0	3.1
$TBOD_5$, mg/L	220	194
TSS, mg/L	200	234
TKN, mg N/L	--	26
NH_3-N, mg/L	--	19

Comparison of these design and actual wastewater characteristics indicate that, on average, the facility has been loaded to approximately its design values. Loadings have exceeded design on a monthly basis.

<u>Operating Results</u>. Design and actual effluent quality are summarized as follows:

Parameter	Design	Actual
$TBOD_5$, mg/L	30	8
TSS, mg/L	30	11
TKN, mg N/L	--	2.4
NO_x-N, mg N/L	--	3.0
TN, mg N/L	--	5.4

These results demonstrate that a high level of performance is achieved at this facility. Effluent $TBOD_5$ and TSS are routinely below secondary treatment levels. An excellent degree of nitrogen control is also achieved. Considering that the facility has been loaded to design values, these results indicate true process capabilities. A probability plot of effluent total nitrogen from this plant is presented in Figure 3-18.

<u>Summary</u>. The Del City WWTP case history demonstrates the performance achievable with an SBR. Similar performance levels have been demonstrated at other full-scale SBR facilities.

3.6.2.11 Conclusion

The case histories described above demonstrate the actual implementation of biological nitrogen removal processes in full-scale wastewater treatment plants. Figure 3-18 presents an overall comparison for the case histories discussed.

The results indicate two "bands" of performance data for the various case histories. Single-sludge, single anoxic/aerobic zone systems generally produce effluents with total nitrogen concentrations of 5 to 15 mg N/L. On the other hand, separate-stage systems and single-sludge, dual anoxic/aerobic zone systems produce effluents with total nitrogen concentrations routinely below 4 mg N/L. Variability from one plant to another is noted, even for plants using the same technology. No clear performance advantage for either separate-stage or single-sludge, dual anoxic/aerobic zone system is apparent. Of the case histories evaluated, the best overall performance is observed for the River Oak facility.

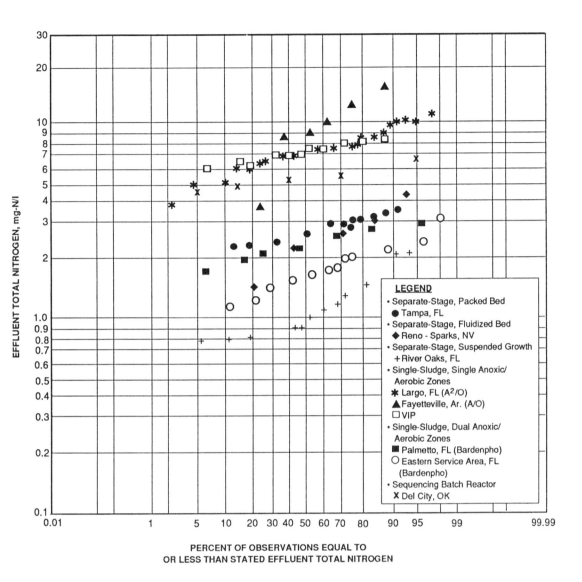

Figure 3-18. Probability plot of monthly average effluent total nitrogen concentrations.

The process selected for a particular plant is dependent on the degree of nitrogen removal necessary. If only a moderate degree of nitrogen removal is required, such as needed to achieve effluent nitrogen concentrations in the range of 8 to 12 mg N/L, a single-sludge, single anoxic/aerobic system should be considered in light of the relatively small additional cost when compared to secondary treatment alone. However, if more complete nitrogen removal is necessary to achieve effluent nitrogen concentrations less than 3 mg/L, the dual anoxic/aerobic zone type of process (Bardenpho) or a separate-stage system should be considered. The additional operating costs (primarily methanol) for the separate-stage system should be seriously considered before that system is selected. In a low alkalinity wastewater, the separate stage system should not be considered due to the alkalinity benefits of the anoxic zone in a single-sludge system.

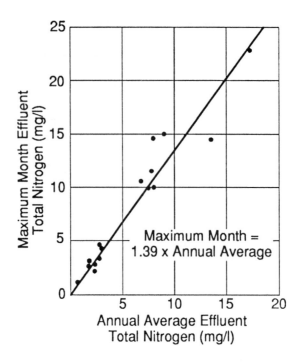

Figure 3-19. Effluent total nitrogen variability for several biological nutrient removal facilities.

Figure 3-19 further quantifies the degree of variability in performance of several biological nutrient removal facilities, including those discussed above(23). Plotted is the highest monthly average effluent total nitrogen concentration for a given year, as a function of the overall average effluent total nitrogen concentration. The results indicate that the monthly maximum is about 40 percent greater than the long-term average effluent quality. Such variations are expected and result from variations in influent wastewater characteristics and process operating conditions. This suggests a relatively small degree of variability compared to other wastewater parameters and suggests that biological nitrogen removal facilities exhibit a high degree of stability.

These types of systems are relatively easy to operate, and they are not prone to upsets or failures. As a result, if an annual average limit on nitrogen is determined to be necessary for a particular receiving water, the stream likely will not receive effluents of widely varying quality on a monthly basis. In other words, the risk to the receiving stream of receiving temporary excessive concentrations of nitrogen from a biological nitrogen removal system is relatively low.

3.7 References

1. U. S. Environmental Protection Agency. **Process Design Manual for Nitrogen Control.** Washington, D.C., 1975.

2. Water Pollution Control Federation. **Nutrient Control.** Manual of Practice FD-7. 1983.

3. Burdick, C. R., D. R. Refling, and H. D. Stensel. Advanced biological treatment to achieve nutrient removal. **Jour. Water Pollut. Control Fed., 54,** 1078, 1982.

4. Ekama, G. A. and G. v. R. Marais. Biological nitrogen removal. In **Theory, Design, and Operation of Nutrient Removal Activated Sludge Processes**. Prepared for the Water Research Commission, P.O. Box 824, Pretoria 0001, South Africa, 1984.

5. Drews, R. J. L. C., and A. M. Greef. Nitrogen elimination by rapid alternation of aerobic/anoxic conditions in orbal activated sludge plants. **Water Res.**, **7**, 1183, 1973.

6. Palis, J. C., and R. L. Irvine. Nitrogen removal in a low-loaded single tank sequencing batch reactor. **Jour. Water Pollut. Control Fed.**, **57**, 82, 1985.

7. van Huyssteen, J. A., J. L. Barnard, and J. Hendriksz. The Olifantsfontein nutrient removal plant. Proceedings of the International Specialized Conference on Upgrading of Wastewater Treatment Plants, Munich, 1989.

8. Metcalf and Eddy, Inc. **Wastewater Engineering: Treatment, Disposal, Reuse**. McGraw Hill Book Company, New York, New York, 1979.

9. Bidstrup, S. M. and C. P. L. Grady, Jr. SSSP--Simulation of single sludge processes. **Jour. Water Pollut. Control Fed.**, **60**, 351, 1988.

10. Grady, C. P. L., Jr. and H. C. Lim. **Biological Wastewater Treatment, Theory and Application**. Marcel Dekker, Inc., New York, New York, 1980.

11. IAWPRC Task Group on Mathematical Modeling for Design and Operation of Biological Wastewater Treatment. **Final Report -- IAWPRC Activated Sludge Model No. 1**. IAWPRC Scientific and Technical Reports, 1986.

12. Grady, C. P. L., Jr., W. Gujer, M. Henze, G. v. R. Marais, and T. Matsuo. A model for single sludge wastewater treatment systems. **Water Sci. Tech.**, **18**, 47, 1986.

13. U.S. Environmental Protection Agency. **Design Manual: Phosphorus Removal**. EPA/62-1/1-87/001, 1987.

14. U.S. Environmental Protection Agency. **Innovative and Alternative Technology Assessment Manual**. EPA/430/9-78-009, 1980.

15. CH2M HILL, **Final Report on the POTW Nutrient Removal Retrofit Study**. Prepared for the Commonwealth of Virginia State Water Control Board, 1989.

16. Kunihiro, C. I., R. D. Reardon, K. C. Wood, W. N. Puhl, Jr., and R. C. Clinger. Feasibility of retrofitting Maryland POTW's for biological nutrient removal. Presented at the 63rd Annual Conference of the Water Pollution Control Federation, Washington, D.C., 1990.

17. Sanitaire Water Pollution Control Corporation. Diffused aeration seminar. Presented to CH2M HILL, Denver, Colorado, 1987.

18. Jenkins, D., M. G. Richard, and G. T. Daigger. **Manual on the Causes and Control of Activated Sludge Bulking and Foaming**. Water Research Commission. Republic of South Africa, 1984.

19. Ingersol-Rand. **Cameron Hydraulic Data**. Woodcliff Lake, New Jersey, 1981.

20. McDonald, D. V. Denitrification by an expanded bed biofilm reactor. **Res. Jour. Water Pollut. Fed.**, **62**, 796, 1990.

21. CH2M HILL. **Fayetteville Pilot Plant Study Final Report.** Prepared for the City of Fayetteville, Arkansas, 1986.

22. Daigger, G. T., G. D. Waltrip, E. D. Romm, and L. M. Morales. Enhanced secondary treatment incorporating biological nutrient removal. **Jour. Water Pollut. Control Fed.**, **60**, 1833, 1988.

23. CH2M HILL. **Biological Nutrient Removal Study.** Presented to the Virginia State Water Control Board, 1988.

Chapter 4

Principles of Chemical Phosphate Removal

4.1 Sources of Phosphorus in Wastewater

Phosphorus occurs in wastewater solely as various forms of phosphate. The types of phosphate present typically are categorized according to physical characteristics into dissolved and particulate fractions (usually on the basis of filtration through a 0.45 micron membrane filter) and chemically into orthophosphate, condensed phosphate, and organic phosphate fractions (usually on the basis of acid hydrolysis and digestion). Table 4-1 presents a summary of this categorization together with examples of typical concentration ranges in U.S.A. municipal wastewaters where no regulations exist on the phosphorus content of synthetic detergents.

Phosphorus originates in wastewater from the following sources: (i) the carriage water (usually minor), (ii) fecal and waste materials, (iii) industrial and commercial uses and (iv) synthetic detergents and household cleaning products. The approximate current per capita contributions of the major sources of phosphate to municipal wastewater in the U.S.A. can be estimated to be: human waste, 0.6 kg P/capita/yr(1); laundry detergents (no product phosphorus limitation), 0.3 kg P/capita/yr(2); and other household detergents and cleaners, 0.1 kg P/capita/yr(3). Industrial, institutional, and commercial sources of phosphorus are highly variable. As such, exact estimates of the amount of phosphate entering a treatment plant must be based on local measurements of the sewage.

Table 4-1. Chemical forms of phosphate in U.S.A. sewage.

Phosphate Form	Typical Concentrations mg P/L
Orthophosphate	3-4
Condensed Phosphates(e.g., pyrophosphate, tripolyphosphate, trimetaphosphate)	2-3
Organic Phosphates(e.g., sugar phosphates, phospholipids,nucleotides)	1

The concentrations of phosphate in U.S.A. municipal wastewaters have, in general, been falling over the past decade. In the late 1960's typical raw sewage total phosphate concentrations were 10-12 mg P/L. Currently, concentrations are usually in the range 3-7 mg P/L in areas where detergent phosphorus content is not regulated. For example, a survey of 11 Virginia and Maryland plants for the period 1982-83 showed an average (flow-weighted) total phosphate concentration of 6.2 mg P/L(4), total phosphate measurements at two North Carolina plants over the period 1984-86 averaged 7.3 mg P/L(5), and the average (flow-weighted) of total phosphate measurements reported for nine Ohio wastewater treatment plants included in a recent survey is 3.0 mg P/L(6).

The reason for this decrease in raw sewage phosphate concentration is most strongly related to changes in household synthetic detergent phosphorus concentration and product type usage. For example, the average phosphorus content of laundry detergents in areas not affected by regulations on phosphorus content fell from about 10.8% P in 1970 to approximately 4.5% P by 1982(3). This decrease has been due to both a decline in the phosphate content of powdered detergents and a significant increase in consumer use of liquid laundry detergents, which do not contain phosphate. In addition, industrial and commercial pretreatment programs removing phosphate before discharge into municipal sewers have contributed to some extent to the reduction.

4.2 Overview of Available Chemical Phosphate Removal Options

Phosphate removal from wastewater involves the incorporation of phosphate into a particulate form (suspended solids) and then the removal of the suspended solids. The types of suspended solids into which phosphate can be incorporated are either biological (micro-organisms) or chemical (sparingly soluble metal phosphate precipitates). The physical removal and subsequent processing of these phosphate-containing solids should be accomplished without allowing significant release of phosphate into liquid streams that are recycled back to the wastewater stream.

Chemical precipitation of phosphate usually becomes necessary when the phosphorus discharge criteria are lower than those that can be achieved by primary sedimentation and secondary biological wastewater treatment. Very few instances of chemical precipitation without the involvement of a biological process exist.

When treating a municipal sewage of average organic strength (BOD_5 = 200 mg/L) to secondary effluent criteria (BOD_5 ≤30 mg/L; TSS ≤30 mg/L) the primary sedimentation and conventional secondary biological wastewater treatment train (i.e., secondary treatment processes that do not incorporate enhanced biological phosphate removal) can remove a maximum of about 2 mg P/L. This performance is dictated by the facts that removal by primary sedimentation is approximately 10%, volatile suspended solids (VSS) in the activated sludge contain about 2.3% phosphorus, a typical standard rate activated sludge has solids that are approximately 80% volatile, and a typical municipal sewage BOD_5-based TSS yield is 0.7.

For example, for a raw influent BOD_5 of 200 mg/L with 30% BOD_5 removal by primary sedimentation, the amount of waste activated sludge produced is (1.0 - 0.3) (0.7 mg TSS/mg BOD_5) (200 mg BOD_5/L) (0.8 mg VSS/mg TSS) = 78 mg VSS/L which contains (78) (0.023) = 1.8 mg P/L. Assuming a maximum of 1 mg P/L removal by primary sedimentation, the total phosphate removed amounts to 2.8 mg P/L. Figure 4-1 illustrates that when the primary sludge and the waste activated sludge are treated by anaerobic digestion there will be some recycle of phosphate from the digestion or the solids handling processes back to the wastewater to be treated. Because of this, the overall phosphate removal is reduced to about 2 mg P/L.

92

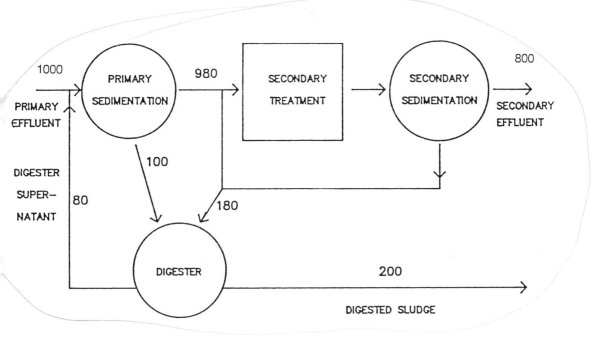

Figure 4-1. Standard rate secondary treatment phosphate removal. Figures are given in kg/day.

Assuming that 2 mg P/L is the limit of phosphate removal achievable by primary sedimentation and conventional biological secondary treatment processes, the effluent phosphorus levels achievable at typical current influent phosphorus levels will be about 4.5 mg P/L (6.5 - 2). Even if detergent phosphate were regulated and the influent phosphorus levels fell by 30%, i.e., to 4.6 mg P/L, the effluent phosphorus concentration would still be about 2.6 mg P/L. There currently are many regions in the U.S.A. (and others being contemplated) where effluent total phosphorus concentrations of 2 mg P/L or less are required (Table 4-2). To achieve these effluent phosphorus concentrations, processes additional to, or other than, conventional biological treatment must be employed. This chapter will concentrate on chemical processes.

Table 4-2. Examples of effluent total phosphate standards (mg P/L).

USA	Great Lakes	1.0 (if >1 MGD)
	Florida	1.0 (lake, bay, impoundment, or estuary discharges)
	Chesapeake Bay Basin:	
	PA (lower Susquehanna)	2.0
	MD	0.2, 1.0, 2.0
	VA (lower Potomac River)	0.18, 0.2, 0.4, 0.5, 1.0
	Washington, DC	0.18, 0.23
	Reno-Sparks, NV	0.4
	Lake Tahoe, CA	1.0
	Tualatin River, OR	0.10, 0.07
Switzerland		1.0 or 85% removal for discharges to lakes
Sweden		≤ 1.0

93

Chemical processes for phosphate removal commonly rely on the formation of sparingly soluble orthophosphates that can be removed by solids separation processes either together with raw sludge and/or waste biological solids or separately. Phosphate precipitation processes can be classified according to their location in the process stream. *Pre-precipitation* refers to the addition of chemicals to the raw wastewater and removal of the formed precipitates together with the primary sludge. *Simultaneous precipitation* refers to the addition of chemicals so that the formed precipitates are removed together with the waste biological sludge. Points of chemical addition that accomplish this are (i) to the primary effluent, and, in an activated sludge plant, (ii) to the mixed liquor, either in the aeration basin itself or to the mixed liquor following aeration but prior to secondary sedimentation. *Post-precipitation* is the addition of chemicals at a point after both the primary and secondary treatment processes. The formed precipitates are removed by an additional solids separation device such as an additional clarifier or a filter.

Phosphate precipitation is achieved by the addition of the salts of one of three metals that form sparingly soluble phosphates. These are calcium (Ca(II)), iron (either ferric iron, Fe(III), or ferrous iron, Fe(II)), and aluminum (Al(III)). The salts most commonly employed are lime ($Ca(OH)_2$), alum ($Al_2(SO_4)_3 \cdot 18 H_2O$), sodium aluminate ($NaAlO_2$), ferric chloride ($FeCl_3$), ferric sulfate ($Fe_2(SO_4)_3$), ferrous sulfate ($FeSO_4$) and ferrous chloride ($FeCl_2$). Pickle liquor, a waste product of the steel industry, containing ferrous iron in either a sulfuric or hydrochloric acid solution, also is used as a source of iron for phosphate precipitation.

A knowledge of the nature of the phosphates formed by addition of these precipitants to wastewater and of their solubilities and the variation of solubilities with solution conditions is essential for predicting and controlling the results of chemical phosphate removal. A list of some of the solids that can form are presented in Table 4-3. In all cases it should be noted that solids other than those containing phosphate can also form. If such solids form, they represent a consumption of dosed chemical and a production of sludge additional to that required for the removal of phosphate.

Table 4-3. Precipitates formed during phosphate precipitation.

Phosphate Precipitant	Precipitates That May Form
Ca(II)	Various calcium phosphates eg. ß-tricalcium phosphate: $Ca_3(PO_4)_2(s)$ hydroxyapatite: $Ca_5(OH)(PO_4)_3(s)$ dicalcium phosphate: $CaHPO_4(s)$ calcium carbonate: $CaCO_3(s)$
Fe(II)	ferrous phosphate: $Fe_3(PO_4)_2(s)$ ferric phosphate: $Fe_x(OH)_y(PO_4)_3(s)$[a] ferrous hydroxide: $Fe(OH)_2(s)$ ferric hydroxide: $Fe(OH)_3(s)$[a]
Fe(III)	ferric phosphate: $Fe_x(OH)_y(PO_4)_z(s)$ ferric hydroxide: $Fe(OH)_3(s)$
Al(III)	aluminum phosphate: $Al_x(OH)_y(PO_4)_3(s)$ aluminum hydroxide: $Al(OH)_3(s)$

[a]Formed by oxidation of Fe(II) to Fe(III) during the treatment process

94

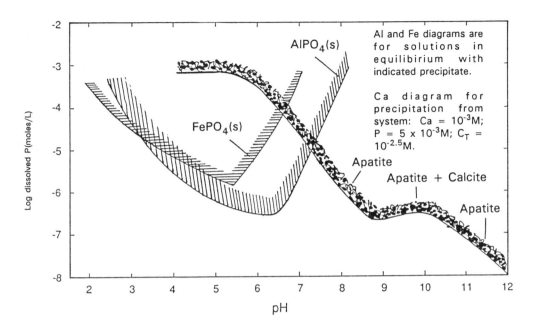

Figure 4-2. Equilibrium solubility diagrams for Fe, Al, and Ca phosphates(7).

The solubility diagrams in Figure 4-2 are illustrations of the commonly accepted variations of the solubility of various metal-phosphate solids with pH. Later in this paper it will be shown that, at least for ferric phosphate, the nature of the solubility versus pH curve may be somewhat different from that depicted in Figure 4-2.

4.2.1 Lime

The solubility curve for "calcium phosphate" shown in Figure 4-2 is one of many that could have been constructed for the wide variety of possible calcium phosphate solids. The calcium phosphate solubility curve presented in Figure 4-2 is for the solids $CaCO_3(s)$ (calcite) and $Ca_5(OH)(PO_4)_3(s)$ (hydroxyapatite). This solubility curve suggests that, to achieve low soluble orthophosphate residuals, the pH must be adjusted to high values (i.e.,pH ≥ 10). This is indeed largely borne out by practical experience with the lime precipitation of phosphate from wastewater. pH values of ≤ 10.5 are commonly used to achieve low phosphate residuals. Because, at these values, the bicarbonate alkalinity of the wastewater will react with the lime as follows:

$$Ca(OH)_2 + HCO_3^- = CaCO_3(s) + H_2O$$

and because this "alkalinity" demand for lime is usually orders of magnitude greater than the lime required for calcium phosphate precipitation, the lime dose for calcium phosphate precipitation is largely determined by the total alkalinity of the wastewater. Lime doses to achieve phosphate removal are equal to approximately 1.5 times total alkalinity (expressed as mg $CaCO_3/L$).

Because the reaction of lime with bicarbonate alkalinity produces calcium carbonate solids, sludge production is also largely related to the alkalinity of the wastewater rather than to the amount of phosphate removed. In the Phostrip process, lime is added to precipitate phosphate that has been anaerobically stripped from a portion of the return activated sludge stream. The activated sludge is rich in phosphorus due to enhanced biological phosphate removal. Under anaerobic conditions the phosphorus is released, producing very high soluble phosphate concentrations. Since the phosphate/total alkalinity ratio of this stream is higher than in the wastewater itself, and lime dose is determined by alkalinity, a greater amount of phosphate is precipitated per unit amount of lime added than if the lime had been added directly to the wastewater. Furthermore, lime treatment of "stripper" effluent often is carried out at pH values of about 9.5 because a higher phosphate residual can be tolerated on this high phosphate concentration stream. The lower required pH value further contributes to lime savings.

Figure 4-2 suggests that low phosphate residuals can be obtained with calcium addition at pH values close to pH 9. This has indeed been demonstrated by Ferguson et al.(7). However, somewhat specialized conditions are required (e.g., Ca(II)/Mg(II) mole ratio $\leq 5/1$) so that the use of slightly alkaline pH precipitation of phosphate is not widely practiced.

Because pH values of ≥ 10 are usually employed for the lime precipitation of phosphate from wastewater, this method cannot be used as a simultaneous precipitation process. The pH values are too high to allow concurrent biological growth. Therefore, lime addition is practiced as a pre- or post-precipitation process only. Raw wastewater treated with lime for phosphate removal may require pH adjustment prior to biological treatment. However, it is possible to effect some reduction in the pH of lime-treated primary effluent by the carbon dioxide produced in biological treatment. Furthermore, nitrification in the activated sludge process will also aid in lowering primary effluent pH. Horskotte et al.(8) devised the so-called ATTF process in which lime precipitation of raw sewage was followed by a nitrifying activated sludge process in which no pH adjustment was required. The aeration basin pH was in the range 7.3 - 8.7 with a primary effluent pH of 11.5.

When used as a post-precipitation process, pH adjustment is required following lime treatment to bring the effluent to within commonly stated discharge limits (pH 6-9) and for the prevention of scaling in downstream processes (e.g., filtration). This pH adjustment is usually achieved by recarbonation followed by clarification to remove the $CaCO_3(s)$ that forms in this process.

4.2.2 Iron and Aluminum

The graphs for $FePO_4(s)$ and $AlPO_4(s)$ solubilities in Figure 4-2 are of similar form. Both show minimum solubilities close to the physiological pH range (6-8.5), i.e., pH approximately 5.5 for $FePO_4(s)$ and pH approximately 6.5 for $AlPO_4(s)$. The minimum solubility of $AlPO_4(s)$ appears to be lower than for $FePO_4(s)$. These two curves were developed for precipitates formed by the addition of reagent grade chemicals to distilled water in the laboratory(9). Discussion later in this chapter shows that the solubility curves observed during simultaneous precipitation of ferric or aluminum phosphates in activated sludge systems are somewhat different. However, it can be stated that low phosphate residuals should be possible by adding either iron or aluminum salts to activated sludge (simultaneous precipitation) as well as by pre- and post-precipitation. This is borne out by results in practice.

When either iron or aluminum salts are added to wastewater to precipitate phosphate, a chemical dose versus soluble orthophosphate residual curve like that shown in Figure 4-3 is obtained. This curve is typical of moderate pH (<7.5) and moderate or low alkalinity (approximately 100 mg $CaCO_3$/L) wastewaters. It indicates that low residual orthophosphate concentrations can be achieved, but only at high Fe(III) doses. This pattern is supported by actual operating data from various plants in the Chesapeake Bay area, presented in Figure 4-4(6). Two predominant regions can be identified -- a "stoichiometric" region at relatively high effluent phosphorus concentrations and an "equilibrium"

96

region at low effluent phosphorus concentrations, with a slight transition between the two regions. In the stoichiometric region the removal of soluble orthophosphate is proportional to (or stoichiometric with) the addition of metal salt. In the equilibrium region, much higher increments of chemical dose are required to remove a given amount of soluble orthophosphate.

Both of these curves can be predicted using an equilibrium model in which one or two precipitates form. For the addition of either of the metal ions Al(III) or Fe(III), the two possible precipitates are a ferric or aluminum phosphate and a ferric or aluminum hydroxide. For a given metal, the formation of these precipitates is dictated by the equilibrium constants governing their solubilities and by the initial pH, alkalinity, and soluble orthophosphate concentration of the sewage.

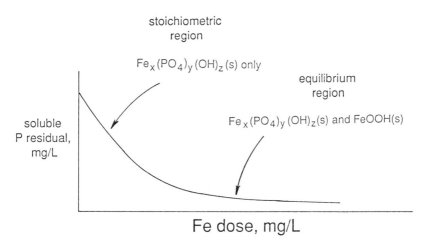

Figure 4-3. Typical Fe dose versus soluble P residual curve.

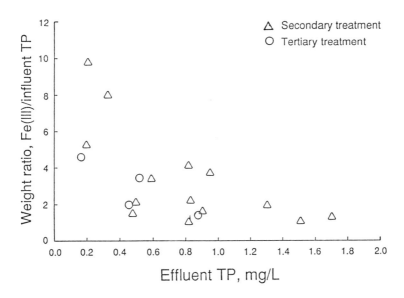

Figure 4-4. Fe(III) to influent TP ratio versus effluent total phosphorus concentration(6).

97

The equilibrium equations used to construct the model for Fe(III) or Al(III) addition are presented in Table 4-4. The most important equation is the one that describes the formation of the metal phosphate precipitate. The actual composition of this precipitate is not known, but most experimental work(9,10,11) suggests that it deviates from the simple forms, $FePO_4(s)$ and $AlPO_4(s)$. An empirical formula for the precipitate with the form $Me_r \cdot H_2PO_4(OH)_{3r-1}$ is widely accepted. The precipitation can be described as:

$$Me_r \cdot H_2PO_4(OH)_{3r-1}(s) = r\ Me^{3+} + H_2PO_4^- + (3r-1)\ OH^- \qquad (1)$$

This formula does not incorporate cations other than Fe^{3+} or Al^{3+} (such as Ca^{2+} or Fe^{2+}) although they may play some role in the precipitation process. Some controversy exists concerning the value of the stoichiometric coefficient r. Recht and Ghassemi(9) estimated $r = 1.2$ moles/mole at pH = 5. Kavanaugh et al.(12) assumed $r = 1$, although their data suggest that r is between 2 and 4. In our experimental work the observed values of r were 0.8 for Al(III) and 1.6 for Fe(III)(13,14). The variability of observed values of r can be explained to some extent by considering the phenomenon of adsorption of phosphate ions on to the precipitate.

Precipitation of a metal hydroxide, MeOOH(s), is also included in the model:

$$Me^{3+} + 2\ H_2O = am\text{-}MeOOH(s) + 3\ H^+ \qquad (2)$$

together with hydrolysis of the metal ion (Me^{2+}) and formation of its hydroxy complexes:

$$Me^{3+} + H_2O = MeOH^{2+} + H^+ \qquad (3)$$

$$Me^{3+} + 2\ H_2O = Me(OH)_2^+ + 2\ H^+ \qquad (4)$$

$$Me^{3+} + 3\ H_2O = Me(OH)_3^0(aq) + 3H^+ \qquad (5)$$

$$Me^{3+} + 4\ H_2O = Me(OH)_4^- + 4\ H^+ \qquad (6)$$

Additional equations constituting the model represent dissociation of phosphoric acid:

$$H_3PO_4 = H^+ + H_2PO_4^- \qquad (7)$$

$$H_2PO_4^- = H^+ + HPO_4^{2-} \qquad (8)$$

$$HPO_4^{2-} = H^+ + PO_4^{3-} \qquad (9)$$

and the formation of soluble complexes of the metal ions with HPO_4^{2-} and $H_2PO_4^-$:

$$Me^{3+} + HPO_4^{2-} = MeHPO_4^+ \qquad (10)$$

$$Me^{3+} + H_2PO_4^- = MeH_2PO_4^{2+} \qquad (11)$$

These soluble complexes, the existence of which is widely accepted (14,15), are responsible for the increase of residual soluble phosphate concentration on the acidic side of the solubility minimum.

If the pH of a system is not controlled, all of these equations must be coupled with equations describing the transformations of carbonate and bicarbonate in a system, either closed or open to the atmosphere.

Table 4-4. Model chemical equations and equilibria constants.

Reaction	pK	
	Fe(III)	Al(III)
$Me_rH_2PO_4(OH)_{3r-1}(s)$ = $r\,Me^{3+}$ + $H_2PO_4^-$ + $(3r-1)\,OH^-$	67.2	25.8
with r =	1.6	0.8
$MeOOH(s)$ + $3\,H^+$ = Me^{3+} + $2\,H_2O$	0.5	-9.1
Me^{3+} + H_2O = $MeOH^{2+}$ + H^+	2.2	4.97
Me^{3+} + $2\,H_2O$ = $Me(OH)_2^+$ + $2\,H^+$	5.7	9.3
Me^{3+} + $3\,H_2O$ = $Me(OH)_3^0(aq)$ + $3\,H^+$	12.0	15
Me^{3+} + $4\,H_2O$ = $Me(OH)_4^-$ + $4\,H^+$	21.6	23
Me^{3+} + $H_2PO_4^-$ = $MeH_2PO_4^{2+}$	-21.8	na
Me^{3+} + HPO_4^{2-} = $MeHPO_4^+$	-9.0	-12.1
H_3PO_4 = H^+ + $H_2PO_4^-$	2.1	
$H_2PO_4^-$ = H^+ + HPO_4^{2-}	7.2	
HPO_4^{2-} = H^+ + PO_4^{3-}	12.2	

All equations of the model and their equilibrium constants are generally well established(15,16,17, 18) with the exception of the equations for the precipitation of metal phosphate and the formation of the metal phosphate complexes. The values of these constants were estimated from experimental data as follows. For large doses of Fe(III) or Al(III), precipitation of the two solids, $Me_rH_2PO_4(OH)_{3r-1}(s)$ and MeOOH(s) will occur. In this situation the residual soluble phosphate concentration is uniquely determined by the pH and any further addition of metal ion should not change its value unless the pH is also changed. Metal ion in excess of that required to precipitate metal phosphate will precipitate as metal hydroxide. The amount of precipitated metal phosphate will correspond to the difference between the initial phosphate concentration and its solubility limit (line AB in Figure 4-5). At decreasing doses of metal ion, two precipitates will still form (that have smaller and smaller metal hydroxide contents) until the dose of metal ion corresponds exactly to the difference between the initial phosphate concentration and its solubility limit. At this point no metal hydroxide will form and all the precipitate will be $Me_rH_2PO_4(OH)_{3r-1}(s)$. For even smaller metal doses only $Me_rH_2PO_4(OH)_{3r-1}(s)$ will precipitate. Thus, at a controlled pH, an initial metal dose will result in a stoichiometric precipitation of metal phosphate characterized by a constant $Me_{added}/P_{removed}$ ratio (stoichiometric region) until two precipitates are formed, at which point the $Me_{added}/P_{removed}$ ratio will increase.

A plot of $Me_{added}/P_{removed}$ versus residual phosphate concentration should look like the inset in the lower part of Figure 4-5. Plots of this type are presented in Figures 4-6 and 4-7 for our pilot plant experimental data on simultaneous precipitation of both ferric and aluminum phosphate and in Figure 4-8 for the ferric phosphate precipitation data from the Blue Plains Wastewater Treatment Plant in Washington, DC (monthly average pH = 6.5-7.1). These plots indicate that the chemical model of metal phosphate precipitation is generally valid. In particular, the existence of two precipitation regions (stoichiometric or one precipitate region and equilibrium or two precipitate region) is evident. The intersection of the horizontal part of the plot with the vertical axis provides an estimate of r and the intersection of the vertical branch with the horizontal axis an estimate of the phosphate solubility.

99

Figures 4-6 and 4-7, and to a lesser degree Figure 4-8 (due to larger scatter), suggest that the values of the $Me_{added}/P_{removed}$ ratio in the one-precipitate region increase slightly with increasing metal doses (decreasing $C_{p,residual}$). In the two-precipitate region it appears that increasing metal doses bring about a small but noticeable decrease of $C_{p,\,residual}$ below the apparent solubility limit. To reconcile both of these features with the chemical model, we postulate that an additional adsorption of phosphate on to the precipitate takes place and that the equilibrium concentration, $C_{p,eq}$, calculated from the model is composed of two parts

$$C_{p,eq} = C_{p,residual} + C_{p,adsorption}$$

or

$$C_{p,residual} = C_{p,eq} - C_{p,adsorption}$$

Only the $C_{p,res}$ part is normally measured as soluble orthophosphate. This adsorption mechanism has been incorporated into the chemical model discussed above (13,14). The adsorption of phosphates on to the formed precipitates causes the observed $Me_{added}/P_{removed}$ ratio to deviate from its true stoichiometric value, which can be estimated by extrapolating the horizontal branch in Figure 4-6 to intersect with the vertical axis.

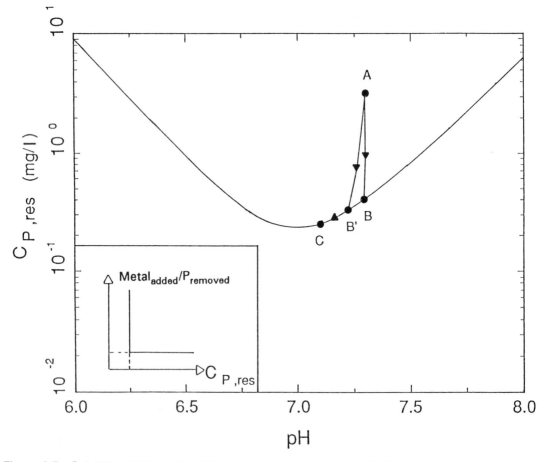

Figure 4-5. Solubility of $Me_rH_2PO_4(OH)_{3r-1}(s)$ co-precipitated with $MeOOH(s)$.

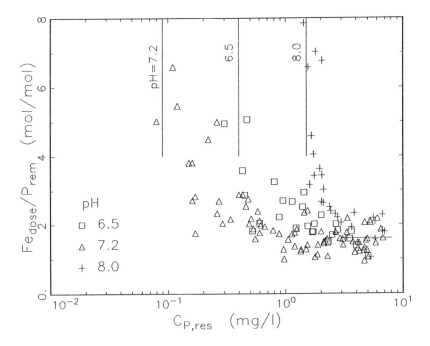

Figure 4-6. Influence of $Fe_{dose}/P_{removed}$ mole ratio and pH on soluble orthophosphate residual. Laboratory pilot plant simultaneous precipitation data (14). (solid line = model prediction)

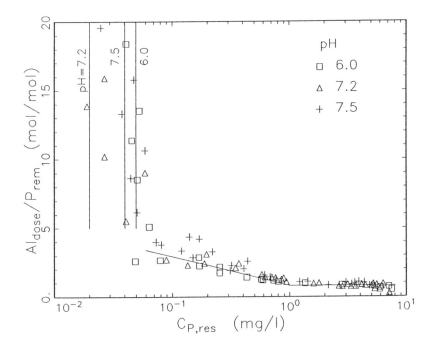

Figure 4-7. Influence of $Al_{dose}/P_{removed}$ mole ratio and pH on soluble orthophosphate residual. Laboratory pilot plant simultaneous precipitation data (13). (solid lines = model prediction)

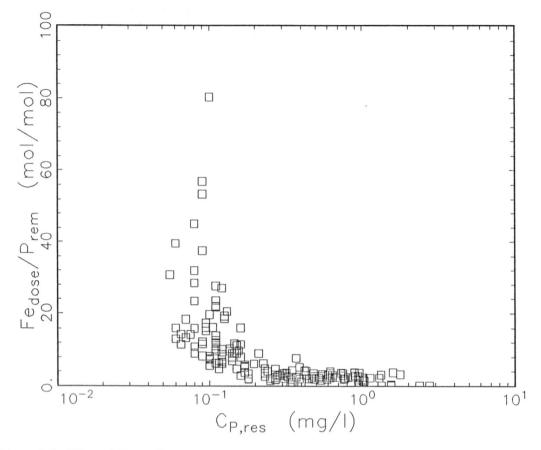

Figure 4-8. Effect of $Fe_{added}/P_{removed}$ ratio on monthly average soluble phosphate residual for data from the Blue Plains Wastewater Treatment Plant, Washington, DC.

The model can be used to calculate the residual soluble orthophosphate concentration as a function of Me(III) dose for specified initial phosphate concentrations and pH values. Calculations of residual soluble orthophosphate concentrations have been performed using parameter values estimated from our experiments(13). An example of the results of such calculations is shown in Figure 4-9 for pH = 7.2 and in Figure 4-10 for results from uncontrolled pH experiments (pH = 7.4-7.8). The points in each of these figures are experimental data ($C_{p,residual}$) corresponding to different initial phosphorus concentrations. The experimental observations agree well with the model predictions. Furthermore, the characteristics of the $C_{p,residual}$-Fe dose curves are similar to those shown in Figure 4-4.

Similar calculations were performed for the conditions at the Blue Plains Wastewater Treatment Plant (using parameter values estimated from results obtained in our laboratory studies (13,14)) and are compared with the observed $C_{p,residual}$ values in Figure 4-11. Taking into account the fact that the Blue Plains data represent monthly averages of soluble phosphate (rather than orthophosphate) resulting from a two-stage addition of both ferric chloride and waste pickle liquor, the agreement between the predicted and observed values is quite good.

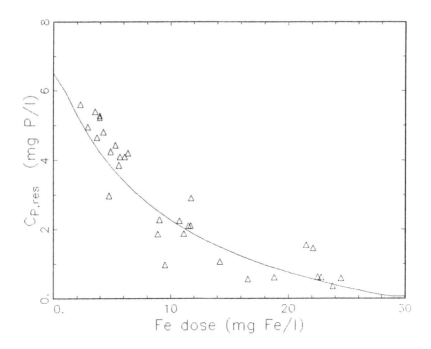

Figure 4-9. Residual phosphate concentration as a function of Fe(III) dose (pH = 7.2, $C_{p,in}$ for observed points = 6-7 mg P/L).

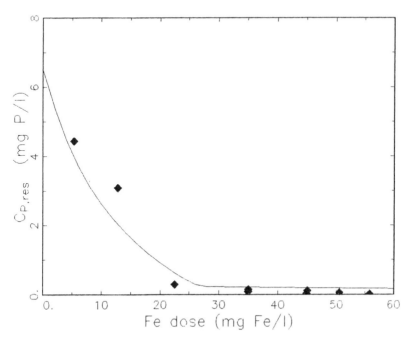

Figure 4-10. Residual phosphate concentration as a function of Fe(III) dose (uncontrolled pH, $C_{p,in}$ for observed points = 6-7 mg P/L).

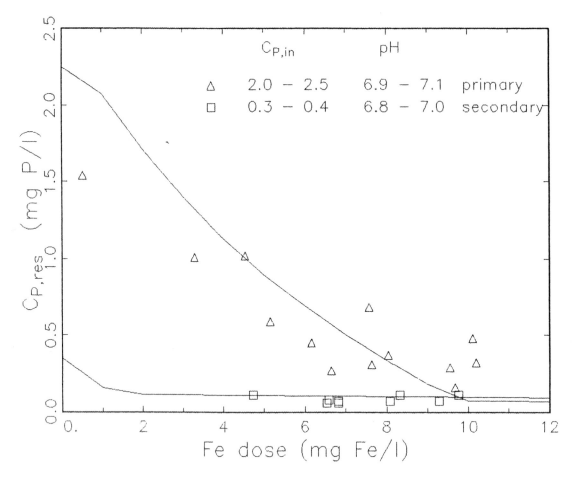

Figure 4-11. Predicted and observed residual phosphate concentrations(model calibrated with experimental data obtained in our laboratory (soluble orthophosphate); observed points from Blue Plains Wastewater Treatment Plant (soluble phosphate)).

To summarize, the mechanisms of phosphate precipitation with metal salts at pH values of less than 8.0 involve the precipitation of $Me_rH_2PO_4(OH)_{3r-1}(s)$ for lower metal salt doses (stoichiometric region) combined with adsorption of phosphate on to the precipitate. Additional phosphate removal due to adsorption results in increasing the observed $Me_{added}/P_{removed}$ ratio as shown in Figure 4-6. When the metal dose is increased and a critical $C_{p,res}$ is reached, $MeOOH(s)$ precipitation occurs resulting in a sharp increase in the $Me_{added}/P_{removed}$ ratio (equilibrium region). The critical $C_{p,res}$ concentration (which is equal to the solubility limit) depends on the pH of the system (Figures 4-12 and 4-13). In the equilibrium region any change of $C_{p,res}$ not associated with a change in pH is caused by phosphate adsorption on metal hydroxide precipitate. The predicted $C_{p,res}$ concentrations are compared with the data of Recht and Ghassemi (9) in Figure 4-14 for Fe(III) addition. This comparison indicates that the residual phosphate concentrations achievable in the pH range of 6.5 to 8 are much lower than those reported by Recht and Ghassemi due to the shift of minimum iron phosphate solubility towards higher pH values. The same observation also applies to simultaneous aluminum phosphate precipitation.

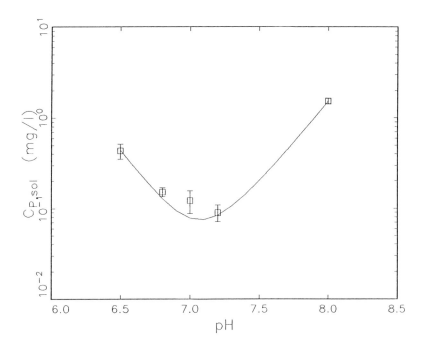

Figure 4-12. Solubility limit for simultaneous precipitation of ferric phosphate. Average experimental data points and ranges of values shown together with model prediction.

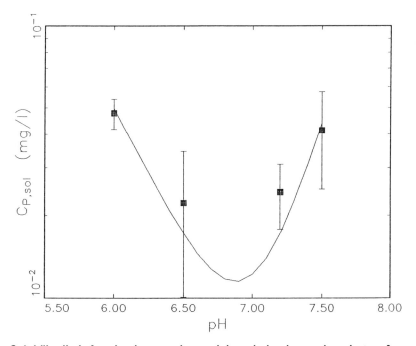

Figure 4-13. Solubility limit for simultaneously precipitated aluminum phosphate. Average experimental data points and ranges of values shown together with model predictions.

105

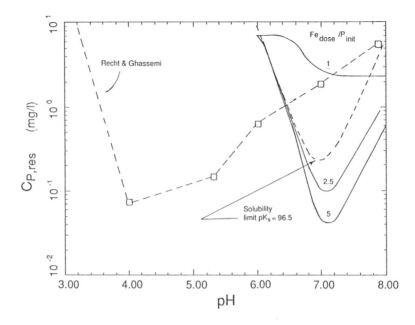

Figure 4-14. Predicted $C_{p, res}$ for various $Fe_{dose}/P_{initial}$ ratios and the data of Recht and Ghassemi(9).

4.3 Sludge Production

The formation of a chemical precipitate by Fe(III) or Al(III) addition for phosphate removal means that there will be an increase in both the mass and volume of sludge produced. Schmidtke (19) has estimated that the average increase of sludge mass and volume upon addition of iron or aluminum salts to a primary plus secondary activated sludge plant to produce a 1 mg total P/L residual is 26% and 35%, respectively. As residual phosphate requirements decrease, the equilibrium precipitation region is reached and either iron or aluminum hydroxides start to form. This additional solids production causes a further increase in sludge production. Data from plants in the Chesapeake Bay area for aluminum or ferric salt addition (Figure 4-15) show that, as the effluent total phosphorus concentration decreases below approximately 1 mg P/L, the sludge generation rates increase significantly(6).

The phosphorus removed both biologically and chemically from wastewater is incorporated into sludge streams which then are subject to a variety of treatment processes. As stated at the beginning of this chapter, to obtain low effluent phosphate residuals one must treat the sludge streams in such a way that the removed phosphate is not returned to the wastewater flow. The critical factors that may cause phosphate release from sludge are changes (usually lowering) in pH and in redox conditions (anoxic or anaerobic conditions rather than aerobic conditions).

In general, these changes do not present a problem for phosphate precipitated by either iron or aluminum salt addition. Aluminum ion does not change its oxidation state over the ranges of redox conditions encountered in sludge treatment. Even though iron(III) is reduced to iron(II) under anaerobic conditions, phosphate is not released because a sparingly soluble ferrous phosphate, $Fe_3(PO_4)_2(s)$ (vivianite) exists. Indeed the addition of Fe(III) to a wastewater stream to precipitate phosphate can result in a decrease in soluble phosphate in the digester supernatant. The difference in pH value between wastewater and a well operated anaerobic digester does not seem to be significant enough to cause phosphate release from phosphate precipitated by iron or aluminum addition during digestion.

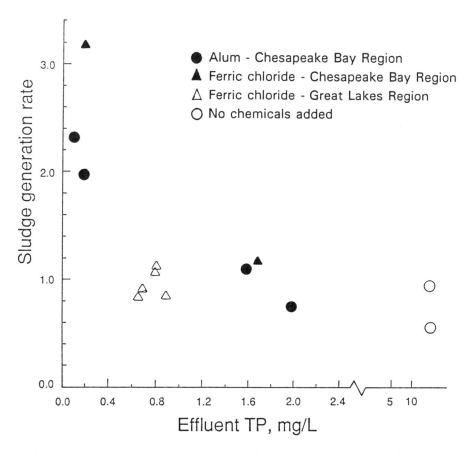

Figure 4-15. Sludge generation rate (total sludge mass/raw TSS) versus effluent TP concentration(6).

Calcium phosphates will only enter a digester when they are produced by lime addition to raw sewage in pre-precipitation processes. Since the pH of a digester is significantly lower than that which will exist in a raw sewage lime pre-precipitation process, phosphate release can be expected. The chemical precipitate produced by post-precipitation is not amenable to anaerobic digestion. Activated sludges originating from systems practicing enhanced biological phosphorus removal leak soluble orthophosphate into solution under anaerobic conditions such as those that exist in anaerobic digestion.

An additional sparingly soluble phosphate-containing solid can form under the conditions encountered in sludge treatment. This is struvite, magnesium ammonium phosphate, $MgNH_4PO_4(s)$. The formation of this material is usually viewed as a nuisance because of its propensity to form scale on the surfaces of sludge and supernatant piping and heat exchangers, and in digested sludge processing units such as vacuum filters, centrifuges and belt presses. Struvite formation is not usually encountered when an anaerobic digester is receiving sludges from a plant where iron and aluminum salts are being used to precipitate phosphate or where Fe(II) or Fe(III) salts are being added to wastewater or the digester for digester gas H_2S control. These cations compete successfully for the phosphate and prevent struvite formation. Indeed, the addition of either Fe(II) (as $FeCl_2$) or Fe(III) (as $Fe_2(SO_4)_3$ or $FeCl_3$) to either wastewater or to an anaerobic digester contents is an accepted method for preventing struvite formation (19).

Based on these observations, struvite formation likely arises from the phosphate released from biological sludges. The digestion of this type of sludge (especially activated sludge) also releases the ammonia (from protein biodegradation) necessary for struvite formation. This being the case, struvite formation should be anticipated from the anaerobic digestion of sludges from plants where enhanced biological phosphorus removal is taking place. This is especially likely since the uptake and release of phosphate from polyphosphate-storing microorganisms is accompanied by the uptake and release of magnesium, the third component of struvite. Data from experiments at Pontiac, MI (21) and York River, VA (22) have suggested that struvite does form from this type of sludge and can act as an "insoluble" sink for some of the biologically removed phosphate. Interestingly, it has been claimed that the struvite formed from such sludges does not cause scaling problems(22); rather, it is claimed that the precipitate forms on the surface of the microbial cells, possibly where the local concentrations of Mg^{2+}, PO_4^{3-} and possibly NH_4^+ being produced from the cell interiors are the highest. This observation is certainly not universally applicable, since Shao et al.(23) have found that struvite scaling in the anaerobic digesters at the City of Los Angeles, CA Hyperion plant disappeared when the activated sludge plant was operated in a fashion that eliminated enhanced biological phosphate removal.

4.4 Summary

To reduce the effluent total phosphorus concentrations from conventionally operated primary plus secondary municipal wastewater treatment plants to 2 mg P/L or below, additional or modified processes must be employed, such as chemical addition or enhanced biological phosphate removal. The most common chemical precipitation techniques are the addition of iron or aluminum salts before and/or into and/or following a biological treatment process. The results obtained in practice using such techniques with both ferric iron and aluminum salts can be successfully predicted from an equilibrium chemical model. Phosphate precipitation by iron and aluminum salts is accompanied by an increase in sludge mass and volume; the additional sludge mass and volume increases as the effluent total phosphorus residual requirement decreases. Iron and aluminum phosphate-containing sludges from pre-precipitation and simultaneous precipitation of phosphate can be treated successfully in anaerobic digestion and sludge dewatering processes without release of phosphate back into solution.

4.5 References

1.   Organization for Economic Cooperation and Development (OECD). **Scientific Fundamentals of the Eutrophication of Lakes and Flowing Waters, with Particular Reference to Nitrogen and Phosphorus as Factors in Eutrophication.** Paris, France, 1971.

2.   Booman, K. A., and R. I. Sedlak. Phosphate detergents - a closer look. **Jour. Water Pollut. Control Fed., 58**(12), 1092, 1986.

3.   The Soap and Detergent Association, personal communication.

4.   Lung, W. S. Phosphorus loads discharged from POTWS in the Chesapeake Bay Drainage Basin. Report to the Soap and Detergent Assoc., New York, NY, 1984.

5.   DeFiore, R. S. Phosphorus removal - case studies chemical and biological. Presented at the North Carolina AWWA/WPCA Annual Conference, November 11, Winston-Salem, NC, 1986.

6.   U.S. Environmental Protection Agency. **Handbook: Retrofitting POTWs for Phosphorus Removal in the Chesapeake Bay Drainage Basin**, EPA/625/6-87/017, 1987.

7.   Ferguson, J. F., J. Eastman, and D. Jenkins. Calcium phosphate precipitation at slightly alkaline pH values. **Jour. Water Pollut. Control Fed.**, **45**, 620, 1973.

8.   Horskotte, G. A., D. G. Niles, D. S. Parker, and D. H. Caldwell. Full-scale testing of a water reclamation system. **Jour. Water Pollut. Control Fed.**, **46**, 181, 1974.

9.   Recht, H. S., and M. Ghassemi. Kinetics and Mechanism of precipitation and nature of the precipitate obtained in phosphate removal from wastewater using aluminum(III) and iron(III) salts. Report No. 17010 EKI for Federal Water Quality Administration, 1970.

10.   Hsu, P. H. Complimentary role from iron(III), sulphate and calcium in precipitation of phosphate from solution. **Environmental Lett.**, **5**, 115, 1973.

11.   Arvin, E., and G. Petersen. A general equilibrium model for the precipitation of phosphate with iron and aluminum. **Prog. Water Technology, 12**, 283, 1980.

12.   Kavanaugh, M., V. Krejci, T. Weber, J. Eugster, and P. Roberts. Phosphorus removal by post-precipitation with Fe(III). **Jour. Water Pollut. Control Fed.**, **50**, 216, 1978.

13.   Luedecke, C., S. W. Hermanowicz, and D. Jenkins. Precipitation of ferric phosphate in activated sludge: A chemical model and its verification. **Water Sci. Technol.**, **21**, 352, 1988.

14.   Gates, D. D., Luedecke, C., Hermanowicz, S. W., Jenkins, D. Mechanisms of chemical phosphorus removal in activated sludge with Al(III) and Fe(III). **Proc. 1990 Specialty Conf. Env. Engng.**, ASCE, New York, 322, 1990.

15.   Stumm, W. and J. Morgan. **Aquatic Chemistry.** Wiley-Interscience, New York, 1970.

16.   Hogfeldt, E. **Stability Constants of Metal-Ion Complexes.** IUJPAC Chemical Data Series No. 21, Pergamon Press, Oxford, 1983.

17.   Smith, R. M. and A. E. Martell. **Critical Stability Constants**, Plenum Press, New York, Vol. 4, 1976.

18.   Feitknecht, W. and P. Schindler. **Solubility Constants of Metal Hydroxides and Metal Hydroxide Salts in Aqueous Solutions.** Butterworths, London, 1963.

19.   Schmidtke, N. W. Sludge generation, handling and disposal at phosphorus control facilities. In **Phosphorus Management Strategies for Lakes.** Ann Arbor Science Publishers, Inc., Ann Arbor, Michigan, 1980.

20.   Dezham, P., E. Rosenbloom, and D. Jenkins. Digester gas H_2S control using iron salts. **Jour. Water Pollut. Control Fed.**, **60**, 514, 1988.

21.   Anon. Naturally occurring struvite precipitation in anaerobic digesters. **Environmental Products Update**, Air Products and Chemicals Incorporated, Allentown, Pennsylvania, September 1985.

22. Sen, D., C. W. Randall and W. R. Knocke. The formation of phosphorus precipitates during the anaerobic digestion of high phosphate activated sludge. Report to Air Products and Chemicals Incorporated, Virginia Polytechnic Institute, 1987.

23. Shao, Y-J, J. Crosse, J. Keller, and D. Jenkins. High rate air activated sludge operation at the City of Los Angeles Hyperion wastewater treatment plant. Presented at Sixth IAWPRC Workshop on Large Wastewater Treatment Plant Design and Operation, Prague, Czechoslovakia, August 1991.

Chapter 5

Design and Operation of Chemical

Phosphorus Removal Facilities

5.1 Introduction

This chapter provides an overview on the design and operation of chemical phosphorus removal facilities. Chapter 4 has discussed the principles of chemical phosphorus removal, including sources of phosphorus in wastewater, the chemistry of phosphorus removal, an overview of removal options, and effects on sludge handling processes.

This chapter reviews the specific process options and discusses the rationale used to select the option appropriate for a given application. Process and facility design procedures and facility costs are reviewed to inform the reader of what is involved in designing, constructing and operating the various chemical phosphorus removal facilities. Impacts on sludge handling are discussed, as well as the increased level of process control and operating costs required over conventional secondary treatment facilities. Finally, a discussion of full-scale process experience is presented, referencing general experience rather than specific cases since the process is widely used.

5.2 Process Options

Chemical phosphorus removal from municipal wastewater typically involves: addition of metal salts (aluminum or iron) or lime to wastewater to form insoluble phosphate precipitates, removal of the precipitate from the wastewater, and disposal of the precipitate with the settled sludge. Many process options are available, but the decisions which must be made by the designer to maximize phosphorus removal and minimize capital and operating costs can generally be classified as follows:

o   Selection of the chemical to insolublize the phosphorus

o   Definition of the point (or points) of chemical addition to the wastewater flow stream

Options to consider in making these decisions are discussed below.

111

5.2.1 Chemical Selection

The factors that influence chemical selection are:

- o   Cost
- o   Alkalinity consumption
- o   Quantities of sludge generated
- o   Safety

<u>Aluminum Salts</u>. Aluminum salts added to wastewater for chemical phosphorus removal include aluminum sulfate(alum) and sodium aluminate. The mechanisms for phosphorus removal using aluminum salts are discussed in Chapter 4. In practice, chemical addition rates are typically higher than what would be predicted based on straight stoichiometry. They are determined from evaluations using the specific wastewater or from general design principles developed through experience.
Alum is the most commonly used aluminum salt, based on the significantly lower cost for this source of aluminum. Consequently, it will be the aluminum salt option discussed in this chapter. Alum addition consumes wastewater alkalinity. This can hinder biological treatment systems in low alkalinity wastewater. Alum can be purchased in either dry or liquid form. The form used will depend on transport costs and owner preference.

Sodium aluminate is not used as frequently as alum but is used with low alkalinity wastewaters and will tend to increase pH and alkalinity. Sodium aluminate is available in either liquid or dry form. Storage and dosing facilities are similar to those required for alum.

<u>Iron Salts</u>. Iron salts typically used for phosphorus removal from wastewater are:

- o   Ferric chloride
- o   Ferrous chloride
- o   Ferrous sulfate

Ferric chloride is available as a commercially prepared liquid. Ferrous chloride and ferrous sulfate are available either as commercial products, or as pickle liquor, a by-product of steel manufacturing. Ferrous chloride and ferrous sulfate can be purchased as a dry product or in a liquid form. The mechanisms of phosphorus removal using iron salts are discussed in Chapter 4.

Ferric chloride and pickle liquor are corrosive liquids that require special precautions in handling, storage, and addition to wastewater to avoid serious injury to personnel and rapid and severe damage to concrete and steel. Ferrous sulfate is relatively stable in its dry form but becomes corrosive when wetted or exposed to high humidity. If the source of pickle liquor is relatively stable and free of undesirable impurities, it is the iron salt typically used in chemical phosphorus removal due to its low price. To achieve maximum phosphorus removal, iron in the ferrous form in pickle liquor must be oxidized to the ferric form. Prior to dosing to the primary clarifiers, pickle liquor is often oxidized by addition of chlorine solution. Ferric chloride is used if a reliable supply of pickle liquor is not available. Ferrous sulfate is not as widely used as ferric chloride or pickle liquor, and will not be discussed further.

<u>Lime</u>. Phosphorus is removed by adding lime to either the primary clarifier or in a tertiary treatment unit following secondary treatment. Phosphorus removal with lime is a water softening process, as described in Chapter 4, and, therefore, the lime dose is dependent on wastewater alkalinity rather than phosphorus content(1). Lime-based phosphorus removal systems are used primarily to meet very low effluent total phosphorus limitations on the order of 0.1 mg P/L. The large amount of sludge generated

from a lime addition system makes the process uneconomical for conventional wastewater phosphorus removal requirements.

Lime-based phosphorus removal systems are either single stage, low lime (pH less than 9.5) systems for 1.0 to 2.0 mg P/L effluent total phosphorus limitations or two-stage, high-lime (pH greater than 11.3) systems to achieve effluent total phosphorus concentrations as low as 0.1 mg P/L.. Lime systems for phosphorus removal in primary clarifiers can be attractive if the primary sludge stabilization process uses lime addition. Lime addition facilities for phosphorus removal require a large investment in equipment and high operation and maintenance costs. Consequently it is rarely used in current designs of wastewater phosphorus removal facilities. Lime addition will not be discussed further in this chapter. Metal salt addition is the most commonly used chemical phosphorus removal process and will be the option discussed in the remainder of this chapter.

5.2.2 Dose points

Three specific metal salt dose points are commonly used in wastewater treatment plants: primary clarifiers, secondary clarifiers, and to tertiary treatment systems consisting of chemical clarifiers and/or filters. Multiple dose points using a combination of the above are also used. Metal salts are dosed upstream of primary and secondary clarifiers and the metal-phosphate precipitate is removed with the sludge. Each dose point could achieve effluent total phosphorus concentrations of about 1 mg P/L. Addition of metal salts to upstream tertiary filters or clarifiers can reduce effluent total phosphorus concentrations to less than 0.5 mg P/L. The dosing points for metal salts and the level of phosphorus removal attainable are summarized in Table 5-1 and shown in Figure 5-1.

A secondary treatment facility with an effluent total phosphorus limitation of about 1 mg P/L would typically be designed for primary clarifier and secondary clarifier dosing points to provide maximum operational flexibility for phosphorus removal. Effluent discharge standards significantly less than 1 mg P/L may require tertiary treatment for phosphate and suspended solids removal.

Table 5-1. Dose point issues.

Dose Point	Anticipated Level of Effluent Total P (mg/L)	Issues
Primary Treatment	≥1	Enhances BOD and TSS removal efficiency Efficient chemical usage Reduces phosphate loading on downstream processes May require polymer for flocculation
Secondary Treatment	≥1	Less efficient chemical use Additional inert solids in MLSS Phosphate carryover in effluent TSS
Primary and Secondary Treatment	1 to 0.5	Combines advantages of above Slightly increased cost
Tertiary Treatment	≤0.5	Required to meet stringent standards Significant increased cost

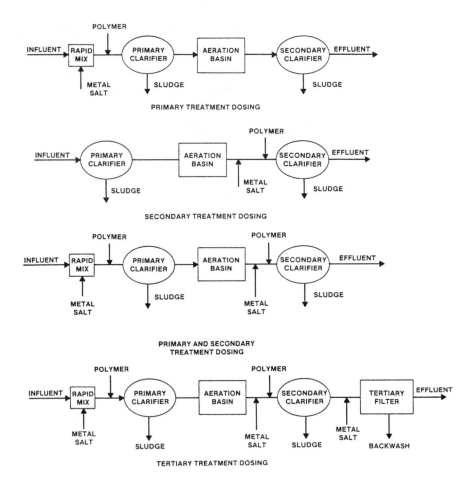

Figure 5-1. Dose points for phosphorus removal.

5.3 Process Selection

5.3.1 Selection Factors

Preliminary evaluation of chemical phosphorus removal processes includes: 1) selection of a metal salt to insolublize the phosphorus, 2) selection of dose points for addition of the metal salt to the wastewater, and 3) estimation of dosage requirements. Chemical dose points will be discussed in detail following this section. It is important to provide for operational flexibility throughout process selection and facility design. The selection procedure must consider all aspects of the phosphorus removal process, including impacts on plant performance, operations requirements, and maintenance needs. Important selection factors are:

o Degree of phosphorus removal required
o Size of WWTP
o Impacts on sludge handling
o Capital cost
o Operation and maintenance cost

o Safety
o Reliability of chemical supply
o Wastewater characteristics
o Skills of WWTP operations personnel

114

As shown in Table 5-1 the metal salt dose points and treatment facilities required are dependent on the level of effluent total phosphorus required. At effluent total phosphorus levels of approximately 1 mg/L, chemical salt addition to conventional secondary treatment processes will be adequate. At lower levels, tertiary treatment facilities (effluent filtration) are required to supplement metal salt addition to remove secondary effluent biological solids containing phosphorus.

Chemical storage and addition facilities for dosage of alum, pickle liquor, or ferric chloride are similar in design and operation. Ferric chloride and pickle liquor are more corrosive than alum. This will increase maintenance costs and require strict safety procedures to prevent injury. For a given wastewater, alum will theoretically produce less sludge than pickle liquor or ferric chloride. For any metal salt, the lower the effluent total phosphorus concentration, the higher the metal salts dose required. Alum sludge can be difficult to thicken and dewater due to entrapment of water in sludge floc and the difficulty of releasing the water from the floc using mechanical processes alone(2,3).

Reliability of chemical supply is a concern for pickle liquor. Pickle liquor is a waste product from steel processing. Consequently, its availability for wastewater phosphorus control is dependent on the location of steel processing facilities near wastewater treatment plants and the production level at those facilities. Although typically less expensive than alum or ferric chloride, pickle liquor may not be available from a reliable source. This should be investigated prior to designing a chemical phosphorus removal facility.

5.3.2 Chemical Comparison

Table 5-2 provides a qualitative comparison of alum, pickle liquor, and ferric chloride for phosphorus removal. In Table 5-2, the plus (+) rating indicates a favorable characteristic or feature of the particular metal salt, and a minus (-) rating indicates an unfavorable characteristic or capability. A zero (0) rating indicates a neutral, or neither positive nor negative characteristic. All ratings are relative to the performance of the three metal salts.

Table 5-2. Metal salts chemical comparison.

	Alum	Pickle Liquor	Ferric Chloride
Process Stoichiometry			
Sludge Production	0	-	-
Alkalinity Consumption	0	-	0
TDS Addition	0	0	0
Operations			
Chemical Efficiency	0	0	0
Ease of Operation	-	+	+
Sludge Characteristics	-	+	+
Contaminants	+	-	+
Maintenance			
Chemical Handling	0	-	-
Corrosion	0	-	-

115

Based on cost, if pickle liquor is reliably available it is typically used to remove phosphorus. However, during initial process evaluations the pickle liquor should be analyzed for contaminants which could harm other treatment processes or cause the concentration of a regulated pollutant in either plant effluent or sludge to exceed the plants discharge permit. If pickle liquor is not available, alum and ferric chloride should be evaluated to determine the effects on the wastewater treatment and solids handling facilities. Alum will typically be used at wastewater treatment plants where solids processing is relatively simple. At facilities employing complex solids handling facilities, the potential thickening, dewatering, and digestion difficulties with alum treated sludge can make alum less attractive for phosphorus removal. At high levels of metal salt addition, solids processing facilities can be adversely affected(2). Since pickle liquor availability is site-specific, the remainder of this chapter will evaluate use of alum or ferric chloride for phosphorus removal.

Process selection is further dependent on the effluent total phosphorus concentration. The ability to dose metal salts to the primary and secondary treatment systems is typically provided for effluent total phosphorus limits down to 1 mg/L. For effluent total phosphorus concentrations significantly below 1 mg/L, filtration of secondary effluent will also be required to remove the particulate phosphorus contained within the biological solids. As illustrated in Chapter 7 (Figure 7-8), the phosphorus contained in suspended solids discharged from a treatment system can contribute significantly to effluent total phosphorus.

5.4 System Design

The facilities required to implement chemical phosphorus removal by metal salts addition consist primarily of chemical storage and feeding equipment. Metal salts are dosed to the wastewater to precipitate phosphorus for subsequent removal. Polymer addition may also be needed to enhance flocculation of the precipitated solids, thereby improving suspended solids removal. In some cases supplemental alkalinity (in the form of lime, sodium hydroxide, or soda ash) is needed to replace alkalinity consumed by metal salts addition.

This section describes the design of chemical storage and handling facilities for chemical phosphorus removal. Chemical addition has other impacts on a wastewater treatment facility. In particular, it can result in a significant increase in sludge production. These impacts are discussed in a subsequent section.

5.4.1 Process Design

5.4.1.1 Chemical Selection

Selection of the metal salt to be used in phosphorus removal will typically be based on predicted or observed performance and cost. Consideration here is focused on the use of either alum or ferric chloride.

Unless specific concerns exist relative to solids processing, alum would often be selected over ferric chloride due to low cost, safety, and corrosion concerns. However, metal salt selection is site specific and is based on the predicted performance for the specific wastewater and on the total cost of the facilities, both capital cost and operation and maintenance cost. Capital costs for facilities to feed the two chemicals will be similar, while operation and maintenance costs (including sludge handling impacts) will vary with wastewater characteristics and effluent total phosphorus limitations.

Using alum, the choice is between dry or liquid chemical for delivery and storage on-site. Liquid alum is more convenient to use, but has higher transportation costs due to shipment of water with the active metal salt. Ferric chloride is typically handled in a liquid form. Design criteria for liquid and dry storage and feeding equipment for alum and ferric chloride are provided elsewhere(1,4).

Regardless of whether alum or ferric chloride is selected or the dosing location, provisions should be made for addition of an anionic polymer to the wastewater to aid solids flocculation. If flocculation between the metal salt and wastewater are less than ideal, pin-floc can develop and result in phosphorus carryover from the primary or secondary clarifier. Addition of polymer can improve solids capture.

Both alum and ferric chloride consume alkalinity, and addition of either of these chemicals can depress the wastewater pH if sufficient alkalinity is not naturally present in the wastewater. Supplemental alkalinity may be added to meet biological treatment requirements or meet effluent discharge permit pH limitations. For wastewater with low alkalinity and a stringent effluent total phosphorus limitation, alkalinity addition facilities can require significant capital and operation expenditures.

A final selection criteria is related to safety of operations personnel. Alum, though corrosive, is easier to handle by operators than ferric chloride or pickle liquor.

5.4.1.2 Range of Doses

Metal salt dose rates should be determined by jar tests or full-scale evaluations of the specific wastewater. Dosage rates will vary depending on influent phosphorus concentration and percent removal desired. The emphasis is on selection of the appropriate range of doses that must be accommodated. The objective in design is to provide maximum flexibility in terms of the types of chemicals, dose points, and range of doses that can be reasonably accommodated. For effluent total phosphorus concentrations greater than 0.5 mg P/L, metal salt dosage will typically vary from 1 to 2 moles of metal salt added per mole of phosphorus removed. At effluent total phosphorus concentrations less than 0.5 mg P/L, metal salt dosage will be significantly higher, approaching values as high as 6 moles per mole of phosphorus removed. On a stoichiometric basis, 9.6 grams of alum are required per gram of phosphorus removed and 5.2 grams of ferric chloride are required per gram of phosphorus removed.

A factor complicating the selection of the chemical dosage is the fact that influent phosphorus concentrations can be quite variable. This variation will affect the metal salt dosage required to maintain effluent total phosphorus concentrations at or below permit limits. Typical system operation is to overdose metal salts at average influent phosphorus concentrations in order to provide adequate concentrations of metals at peak influent phosphorus loads.

Jar tests are useful in comparing typical design dosage values to dosages required to remove phosphorus from a specific wastewater. Jar testing procedures are detailed in other references and will not be described here(5). Results from jar tests must be used with caution, since the tests are intended to simulate, but will most likely not duplicate, full-scale conditions. If possible, tests should be run on a full-scale treatment plant. If the full-scale facility is not available, the jar test results should be used with caution and a capability to dose a range of metal salt dosages should be incorporated into the design. Properties of commercially available ferric chloride and alum and example calculations of dosages are shown in Tables 5-3 and 5-4.

Table 5-3. Ferric chloride dosage (6).

Chemical Formula: $FeCl_3$

Molecular Weight: 162.3 grams/mole

Assume ferric chloride solution @ 30 percent $FeCl_3$ by weight:

> Weight per gallon: 11.2 lb/gal
> $FeCl_3$ per gallon: 3.37 lb/gal

Theoretical Dosage   = 1 mole $FeCl_3$ per mole P = 5.24 lb $FeCl_3$ per lb P

Assume the specific wastewater requires 2 moles $FeCl_3$ per mole P. The dosage of 30 percent ferric chloride solution per pound phosphorus is calculated as follows:

$$\frac{(5.24 \text{ lb } FeCl_3/\text{lb P})}{(1 \text{ mole } FeCl_3/\text{mole P})} \quad \frac{(1 \text{ gal } FeCl_3 \text{ solution})}{3.37 \text{ lb } FeCl_3} \quad \frac{(2 \text{ mole } FeCl_3)}{\text{mole P}} = 3.1 \text{ gal } FeCl_3 \text{ solution/lb P}$$

If WWTP influent TP concentration = 10 mg P/L, dosage of $FeCl_3$ solution per MG influent flow is:

> (10 mg P/L) (8.34) (1 MG) (3.1 gal $FeCl_3$ solution/ lb P) = 258.5 gal $FeCl_3$ solution/MG

Table 5-4. Alum dosage (1,4).

Chemical Formula: $Al_2(SO_4)_3 \cdot 14\ H_2O$

Molecular Weight: 594.3 grams/mole

Assume alum solution @ 49 percent $Al_2(SO_4)_3 \cdot 14\ H_2O$ or 4.37 percent as aluminum.

> Weight per gallon: 11.1 lb/gal
> Aluminum weight per gallon: 0.485 lb/gal

Theoretical Dosage   = 1 mole Al per mole P or 0.5 mole alum per mole P = 0.87 lb Al per lb P

Assume the specific wastewater requires 2 moles Al per mole P. The dosage of 49 percent alum solution per pound phosphorus is calculated as follows:

$$\frac{(0.87 \text{ lb Al/lb P})}{(1 \text{ mole Al/mole P})} \quad \frac{(1 \text{ gal alum solution})}{0.485 \text{ lb Al}} \quad \frac{(2 \text{ mole Al})}{\text{mole P}} = 3.6 \text{ gal Al solution/lb P}$$

If WWTP influent TP concentration = 10 mg P/L, dosage of alum solution per MG influent flow is:

> (10 mg P/L) (8.34) (1 MG) (3.6 gal alum solution/lb P) = 300.2 gal alum solution/MG

Process designers should plan for polymer addition facilities to help coagulate solids and enhance the performance of metal salt addition for phosphorus removal. Anionic polymer dosages to the wastewater will range from 0.1 to 0.5 mg/L. As with metal salt addition, polymer type and dosage should be established for the specific wastewater. Suppliers can be contacted for assistance in selecting a cost-effective polymer.

In summary, a range of dose rates should be established based on treatment objectives and variations in influent waste strength. This range then forms the basis for facility design.

5.4.1.3 Storage Requirements

Alum can be delivered and stored on-site in either a liquid or a dry form. Ferric chloride is typically delivered as a liquid. Polymer is readily available in either liquid or dry form to meet the preference of the operating staff. Liquid chemicals are easier to handle and store, but are more expensive to purchase and transport on a unit cost basis than dry chemicals due to higher shipping costs.

A minimum storage volume should be determined for each chemical. Those minimum volumes should take into account: average day usage, peak day usage, and local supplier delivery schedules and delivery volumes. Based on those factors, the minimum on-site storage volume should be the greater of:

    o   Two weeks consumption at average day use
    o   Three days consumption at peak day use
    o   150 percent of typical delivery volumes

In some cases, a storage volume equal to one month of consumption at average usage is provided, depending on the reliability of supply. Certain chemicals such as liquid polymer can deteriorate during storage. The on-site storage volume should be adjusted, if necessary, to take this into account. Manufacturers should be consulted for information on the storage life of their product.

Care must be taken in on-site storage to assure that the temperature of the chemical is maintained above the point at which it begins to crystallize. A 30 percent ferric chloride solution will freeze at -58°F; alum solutions will begin to crystallize at 30°F. A 50 percent solution of sodium hydroxide should be maintained at a temperature above 55°F to avoid crystallization. Chemical suppliers should be consulted for specific characteristics of the chemical supplied. If the temperature of the storage and chemical metering areas could fall below the desired minimum, supplemental heaters should be provided. Temperature must be maintained, not only in the storage tank, but also in the pipeline conveying the bulk chemical. This can be accomplished by heat tracing and insulation. Humidity control should be provided for dry polymer and dry alum storage areas since both chemicals will absorb moisture from the air.

Using the required molar dosage of metal salt, the anticipated phosphorus concentration, properties of the metal salt as delivered to the plant, and the desired on-site storage capacity, the required storage area or storage volume can be calculated. An example calculation of ferric chloride storage requirements is shown on Table 5-5. Calculations for other chemicals are similar.

Safety of operations personnel should be incorporated by the process designer early in facility development. The chemicals used in phosphorus removal are irritants, corrosive, hazardous, and if mixed with incompatible chemicals can release large quantities of steam and heat. Suppliers and manufacturers of the chemicals should be consulted to develop safety procedures to prevent injury to personnel and deterioration of treatment facilities.

Table 5-5. Example calculation of ferric chloride storage requirements.

From Table 5-3, for 10 mg/L influent P, dosage of 30 percent $FeCl_3$ = 258.5 gal $FeCl_3$ solution/MG

Assuming: Average day WWTP flow = 10 MGD   Peak day WWTP flow = 25 MGD

Storage Required for 30 days storage at average day usage

   = (258.5 gal $FeCl_3$ solution/MG) (10 MGD) (30 days) = 77,550 gallons

Storage required for three days at peak day usage

   = (258.5 gal $FeCl_3$ solution/MG) (25 MGD) (3 day) = 19,390 gallons

Typical delivery volume is 4,000 gallons per truckload. 150 percent of this is 6,000 gallons.

Provide at least 77,550 gallons $FeCl_3$ solution storage

Storage Tanks: Provide three horizontal storage tanks, each 12 feet diameter by 30 feet long. Total storage volume = 79,400 gallons. Average storage life is just over 30 days, which is more than acceptable for $FeCl_3$.

5.4.1.4 Equipment Sizing and Controls

Once the chemical storage requirements have been determined, preliminary sizing of storage tanks or dry chemical bins should be accomplished. Sizing criteria should include a minimum of two storage tanks for a given chemical to provide redundancy and ease of maintenance for critical unit processes.

From required chemical dosages, metering pump capacities can be determined. If dosage volumes are small, diaphragm metering pumps or small progressing cavity pumps should be used. For very large dosages, centrifugal pumps or large progressing cavity pumps may be required. The practical maximum capacity of a diaphragm metering pump is approximately 500 gallons per hour. Pump redundancy or piping interconnection between pumps for similar service should be provided to assure continuous operation of critical processes in the event of pump failure. Pumps used in chemical phosphorus control include: horizontal end suction centrifugal, vertical wet pit centrifugal, progressing cavity, and diaphragm metering. A detailed description of materials of construction will be included in the section on facility design.

Once the number and approximate size of chemical storage tanks and metering pumps have been determined, a preliminary facilities layout should be prepared. Storage tanks and feed pumps should be located as close as possible to the intended dosing points to minimize discharge piping.

Containment areas should be provided for the liquid chemicals which are sized to retain the volume of the largest tank in anticipation of tank rupture or pipe breakage. Ample room should be allotted around the tanks and pumps for maintenance and addition of future units. For large facilities with multiple dose points, smaller day tanks remote from the large storage tanks are often used. Sufficient volume of chemical solution for one day of operation is pumped to each day tank, and the metering pumps are supplied from that tank.

The process designer should determine the strategies necessary to control the chemical phosphorus removal system. To date, automatic phosphorus measurement systems have not proven reliable for full-scale application. Typically, chemical dosing systems are paced on influent flow for systems where influent phosphorus concentration is relatively constant, or the dosage is set based on the anticipated peak phosphorus loading to avoid violation of the discharge permit limitations. For this second scenario, overdosing of metal salts will occur at influent phosphorus conditions less than peak. Both systems rely on periodic testing of influent and effluent phosphorus concentrations and use of that information in combination with wastewater flow rate to develop dosing criteria. Wastewater pH can be measured using on-line systems to control the dose of caustic or other chemical used to replace alkalinity that is consumed by metal salt addition.

5.4.1.5 Dose Points

Schematic diagrams of dose points for metal salt addition were shown in Figure 5-1. A more detailed discussion of specific dose points will be presented in this section. Dose point location can be critical to successful system operation and chemical dosage minimization. Important design parameters are:

- o   Location of chemical addition
- o   Methods of chemical addition
- o   Method of achieving flash mixing of metal salt and wastewater
- o   Development of floc particles
- o   Polymer addition to aid settling of floc

Primary Clarifiers. In addition to precipitation of phosphorus compounds, metal salt addition upstream of the primary clarifiers enhances suspended solids and BOD removals in the primary clarifiers due to coagulation of suspended organic matter. Removal of organic material in the primary clarifier reduces the loading to the secondary treatment facilities, resulting in capital and operation and maintenance cost savings for secondary treatment.

The optimum addition points are as far upstream of the primary clarifier as possible, and to facilities that generate large amounts of turbulence such as: centrifugal pump suction, hydraulic jump in a Parshall flume, flow splitting structure, or aerated grit basin. It is important to disperse the metal salt throughout the wastewater to minimize chemical dosage. Immediate chemical dispersion will also minimize deterioration of concrete and steel in basins and channels where the metal salt is added. To improve distribution, a chemical solution header or multiple injection points may be used. An example of a chemical solution header is shown in Figure 5-2. Dilution water is often added to assist with dispersion of the chemical solution.

The optimum flash mixing method is to pass the wastewater through a rapid mix basin where the metal salt is added and the entire contents of the basin agitated with a high intensity mixer to provide intimate contact. A rapid mix basin is not a typical unit process in a biological wastewater treatment plant. It will often be less costly to overdose metal salts to adjust for the lack of ideal mixing than to construct the mix basins, especially at existing plants. Once the metal salt and wastewater have been mixed, the metal salt and phosphorus will form precipitates. To settle these precipitates in the primary clarifier they must join with other precipitates to form floc particles large enough and with enough mass to settle and be removed with the primary sludge. Gentle agitation of the wastewater promotes inter-particle contact and enhances flocculation. Areas upstream of the primary clarifier that promote flocculation include: aerated or mechanical grit chambers, flow splitting structures, and influent wells of primary clarifiers, whether designed conventionally or as a flocculator.

121

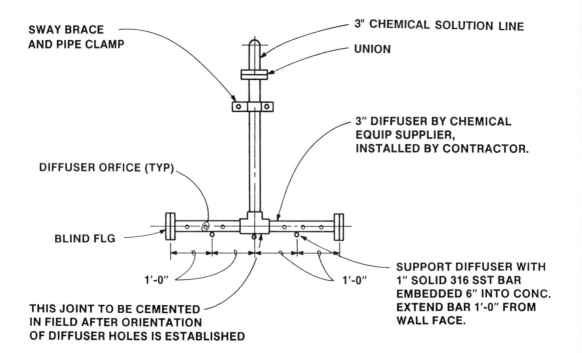

SWAY BRACE
AND PIPE CLAMP

3" CHEMICAL SOLUTION LINE

UNION

3" DIFFUSER BY CHEMICAL
EQUIP SUPPLIER,
INSTALLED BY CONTRACTOR.

DIFFUSER ORFICE (TYP)

BLIND FLG

1'-0"

1'-0"

SUPPORT DIFFUSER WITH
1" SOLID 316 SST BAR
EMBEDDED 6" INTO CONC.
EXTEND BAR 1'-0" FROM
WALL FACE.

THIS JOINT TO BE CEMENTED
IN FIELD AFTER ORIENTATION
OF DIFFUSER HOLES IS ESTABLISHED

Figure 5-2. Chemical solution diffuser.

In some applications, addition of an anionic polymer to the wastewater is required to enhance particle coagulation and floc formation. Polymer is added upstream of the primary clarifier following rapid mixing and flocculation of the metal salt. Polymer should be added as a dilute solution to the wastewater since concentrated solutions require very intense mixing to assure dispersion, which could break up previously formed floc.

Metal salt addition to primary clarifiers alone can be expected to remove 70 to 95 percent of influent phosphorus depending on dose rate(1).

Secondary Clarifiers. Addition of metal salts upstream of the secondary clarifiers provides a high level of phosphorus removal. At this point in the treatment process phosphorus is typically in the orthophosphate form which can be precipitated with the metal salt, or it is included with the biomass. The metal salt and phosphorus precipitate can be removed with the flocculent biomass in the secondary clarifier.

Metal salt addition points are where flash mixing with the wastewater can be best achieved. Turbulence is present in the downstream end of aeration basins, aerated distribution channels, and flow splitting structures. In activated sludge systems intense agitation is not desirable prior to entering the secondary clarifiers due to potential destruction of biological floc which can reduce clarifier efficiency. Overdosing metal salts can compensate for some mixing inefficiencies.

The turbulence points described previously can also provide necessary flocculation to enmesh the coagulated phosphorus particles with the biological floc for removal in the secondary clarifier. If pin-floc develops, polymer addition upstream of the secondary clarifier would aid suspended solids removal.

122

Tertiary Filters. When low effluent phosphorus levels are required, less than 0.5 mg/L total effluent phosphorus, effluent filtration will be necessary. Tertiary treatment facilities would more likely be designed with rapid mixing and flocculation basin facilities to allow for optimum metal salt addition. Polymer addition could be required if solids concentrations are high enough to achieve flocculation in a tertiary clarifier. If suspended solids are low from the secondary treatment system (less than 30 mg/L), secondary effluent with metal salt addition can be applied directly to the tertiary filter without additional clarification or polymer addition. The phosphorus precipitates will be removed in the tertiary filter.

Multiple Dose Points. To offer the WWTP operator maximum flexibility to meet effluent phosphorus limitations at a minimum cost, the designer should provide for multiple metal salt and polymer dose points. Metal salt and polymer addition to the primary clarifiers, secondary clarifiers, and tertiary treatment system (if needed) are decisions made during facility design that will reduce operating costs. The flexibility offered to the operator is use of any or all of the dose points to optimize WWTP performance.

5.4.2 Facility Design

5.4.2.1 Materials

Process design criteria are used to size the various facilities. After the facilities are sized, detailed design drawings and specifications for the physical/chemical phosphorus removal facilities are developed. Facilities must be designed for storage and dosage of metal salts, polymer, and supplemental alkalinity (if necessary).

Once minimum chemical storage volumes have been determined, the volumes can be compared to standard storage tank sizes to lay out the necessary storage facilities. General guidelines would require a minimum of two tanks for each chemical to provide redundancy. If tanks are constructed from fiberglass reinforced plastic (FRP) (which would be acceptable for alum, ferric chloride, pickle liquor, polymer, and caustic) a large number of tank dimensions and configurations are available.

Storage tanks should be constructed within diked chemical containment areas that can hold the contents of the largest tank if ruptured. Extreme caution must be observed in storing different chemicals within a common containment area to assure that the chemicals are compatible in their concentrated forms. For example, mixing concentrated ferric chloride and caustic (sodium hydroxide) will result in a violent reaction generating high temperatures and releasing steam.

Piping needs will include transfer piping from the chemical unloading facilities to the storage tanks, suction piping from the tanks to the chemical metering pumps, and discharge piping from the metering pumps to the point(s) of chemical addition. Piping should be suitable for the solution to be conveyed over the range of anticipated operating temperatures and pressures. PVC piping is suitable for the chemicals used in phosphorus removal. Piping supports and piping expansion provisions must be detailed. Local regulatory codes should be reviewed to determine the need for shields or covers over chemical piping joints and valves to protect personnel from leakage of a pipe under pressure.

System control, shutoff, pressure relief, and check valves must be selected with care for the intended service. Ball and diaphragm valves are used as process control and shutoff valves. Pressure relief valves are often provided with the chemical metering pumps. Two ball check valves in series are used to protect concentrated chemical solutions from contamination when injecting into wastewater flow streams.

123

The pumps used in chemical phosphorus removal were presented earlier in this chapter. Diaphragm metering pumps are typically used to deliver the required dosage of chemicals to the wastewater flow stream. These pumps are reliable, have at least a 10:1 flow rate range, and will accurately provide a consistent flow rate. Double diaphragm metering pumps are recommended due to a higher level of reliability. An isometric view of a metering pump installation is shown in Figure 5-3.

Centrifugal pumps are used to transfer large volumes from a main storage tank to remote day tanks. Pump casing and impeller materials must be suitable for the liquid to be pumped. Manufacturers must be consulted on suitability of their pumps. Vertical wet pit centrifugal pumps are used in containment area sumps to pump out spilled chemicals and washdown water. Locating the motor out of the liquid simplifies pump maintenance. A clean water source is often needed to lubricate the pumps if the pumped material is too corrosive or abrasive. Teflon is often used in these pumps due to high corrosion resistance.

Progressing cavity pumps are used to convey concentrated liquid polymer solutions due to high viscosity. The pump stator and rotor must be resistant to the polymer; Buna-N is often used for the stator with a chromed rotor. Motor horsepower should be carefully selected to compensate for the increased motor loads caused by the high viscosity of polymer solutions. The effects of temperature on polymer viscosity must also be considered when selecting pump motor horsepowers. Using variable frequency drives to adjust pumping speed in order to adjust polymer dose in response to changing waste loads can reduce operating costs.

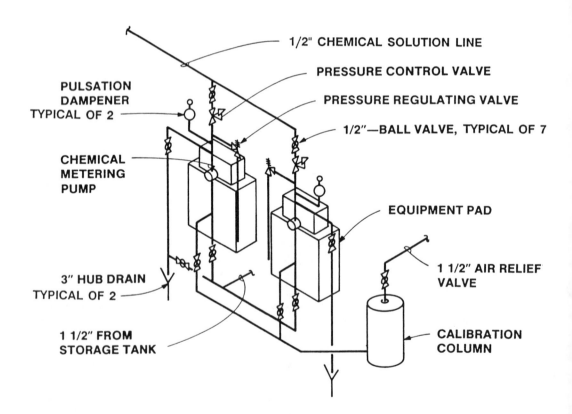

Figure 5-3. Chemical metering pump and piping schematic.

Chemical resistant grating should be installed in chemical containment areas. The grating can keep the operator raised above spilled corrosive and hazardous chemicals and keep slippery polymers off of operator walking surfaces. An elevated grating platform allows installation of chemical piping below grade instead of overhead which would pose a greater potential for exposure to chemicals due to leakage. FRP grating with a nonskid surface is suitable for the chemicals used.

Concrete containment areas should be coated with a corrosion resistant non-slip coating rated for the stored chemicals to prevent discoloration and deterioration of the concrete and steel. Ferric chloride is extremely corrosive and will leave deep orange stains on concrete.

Safety equipment must also be included in the design. Eyewash and safety showers should be easily accessible. A locker stocked with protective gear must be near the storage areas. Safety signs should be prevalent and clearly understood by operating and maintenance personnel. A supply of washdown water should also be provided along with a permanently mounted hose.

5.4.2.2 Controls

The mode of system operation must be determined before the facility controls can be designed. Due to a lack of reliable process monitoring equipment, controls tend to be relatively simple. Control of chemicals for small facilities are either manual, pH paced, flow paced, or adjusted on predicted diurnal loadings. A more detailed discussion of process control will be presented later in this chapter.

5.4.2.3 Facility Costs

Implementation of a chemical phosphorus removal system requires installation of chemical storage, metering, and piping facilities, as described above. Additional sludge handling facilities may also be required.

Several general cost estimating guides are available to develop order of magnitude cost estimates for the construction of the required facilities. The Water Pollution Control Federation Manual of Practice entitled **Nutrient Control**(7) and a recent U.S. EPA manual (4) provide cost curves that may be useful in developing preliminary cost estimates. The U.S. EPA **Innovative and Alternative Technology Assessment Manual**(8) also provides preliminary cost information for a wide variety of facilities, including sludge handling.

Table 5-6 provides order of magnitude costs for various components of a chemical phosphorus removal system. These costs may be used to develop a more definitive cost once a specific system layout is developed. The values listed are for mid-1988 and must be adjusted for inflation to the estimated mid-point of construction of the facility.

The values listed in Table 5-6 are generalized; the actual costs for a particular installation will vary depending on local conditions. However, they may be useful for preliminary project planning. Importantly, an allowance in facility cost estimates for unforeseen items that will almost certainly develop must be incorporated into any preliminary design cost estimate.

Table 5-6. Chemical phosphorus removal facilities costs.

FRP Storage Tanks	3,000 gal: $3,100/EA 6,000 gal: $4,500/EA 12,000 gal: $7,800/EA
PVC Piping--Schedule 80	1 inch dia: $ 9/LF 2 inch dia: $15/LF 4 inch dia: $24/LF
Lined Steel Piping	4 inch dia: $48/LF
Diaphragm Valves (Teflon Lined)	1 inch dia: $140/EA 2 inch dia: $210/EA 4 inch dia: $590/EA
PVC Ball Check Valves	1 inch dia: $140/EA 2 inch dia: $210/EA 4 inch dia: $259/EA
PVC Ball Valves	1 inch dia: $ 25/EA 2 inch dia: $ 50/EA 4 inch dia: $200/EA
Fiberglass Grating	$15/SF
Acid Resistant Concrete Coating	$15/SF
Eyewash and Safety Shower	$465/EA
Duplex Sump Pump	$ 6,000/EA
Diaphragm Metering Pump	$ 8,200/EA
ANSI Rated Centrifugal Pump	$12,000/EA
Progressing Cavity Polymer Pump	$ 9,800/EA
Polymer Mixing/Aging System	$62,000/EA
Building on Grade	$60/SF
Electrical/Instrumentation/Control	15 percent of construction cost

EA = Each
LF = Lineal Foot
SF = Square Foot

5.5 Sludge Handling Impacts

This section will qualitatively discuss the impacts on the various solids handling processes at the WWTP due to chemical removal of phosphorus.

5.5.1 Solids Generation

Additional solids will be generated at wastewater treatment plants when chemical phosphorus removal is used. This factor was discussed briefly in Chapter 4. Actual operating experience at full-scale wastewater treatment plants is summarized elsewhere (1,4,7). Preliminary estimates of the increase in dry solids production rates can be made based on process stoichiometry.

The location of chemical dose points can effect solids production. Metal salt addition to the primary clarifier can result in a 50-100 percent mass increase in primary sludge and an overall increase in plant sludge mass of 60-70 percent. Metal salt addition to the secondary clarifier can increase activated sludge mass by 35-45 percent and overall plant sludge by 10-25 percent(1). However, multiple addition points will result in optimum chemical dosage and reduced sludge production. An increase in sludge volume due to a reduction in the concentration of settled sludge should be anticipated, as discussed in Chapter 4. In fact, optimization of chemical doses and dose points involves minimization of both chemical costs and sludge handling impacts.

5.5.2 Clarification

Primary clarifiers are designed based on a surface overflow rate for readily settleable particles. When the metal ion/phosphate precipitate is added, peak overflow rates must not exceed 1,200 gallons per day per square foot (gpd/sf) to assure good removal(1). Addition of polymer to the primary influent can enhance flocculation of the precipitate into more readily settleable particles, which can increase the allowable overflow rate. Site specific testing is required to quantify allowable increases in peak overflow rates with polymers.

Secondary clarifiers are designed for readily settleable sludge (trickling filters) or poorly settling sludge (activated sludge). Secondary clarifiers are designed based on both hydraulic loading and mass solids loading. Metal salt addition to the secondary clarifiers will increase mass solids loading and could reduce the settling characteristics of the secondary sludge. Polymer addition to the secondary clarifiers should be provided to aid capture of metal/phosphorus particulates that are not enmeshed in the biological floc. The metal salt will increase the nonvolatile solids fraction of activated sludge, resulting in a higher mixed liquor suspended solids concentration to maintain the same mass of active microorganisms in the aeration basins. Consequently, overall plant organic treatment capacity may be negatively affected. The increase in sludge mass and volume will require larger capacity primary sludge, and return and waste activated sludge pumps.

5.5.3 Thickening and Dewatering

There are a wide variety of sludge thickening and dewatering processes currently in use in wastewater treatment plants. Biological wastewater sludge is difficult to thicken and dewater. A large portion of a plant's capital cost and yearly operation and maintenance budget can be devoted to solids processing and disposal. To a large treatment plant in an urban community the additional volume of sludge for disposal can be very costly, while the impact at a small rural facility can be minor.

In addition to the increased mass of solids to be processed, chemical phosphorus removal sludge can be difficult to thicken and dewater. The reasons for the resistance to thickening and dewatering are not well understood, nor are they uniform in degree of difficulty at specific facilities. Alum treated biological sludges are generally more difficult to thicken and dewater than are ferric chloride treated sludges. Due to the site-specific nature of sludge thickening and dewatering, it is recommended that a chemical phosphorus removal evaluation and design include pilot- or full-scale testing of the effects of the sludge on solids thickening and dewatering facilities.

5.5.4 Digestion

Anaerobic digester performance can be affected by metal salts addition to wastewater for phosphorus removal. The effect on digestion is usually a reduction in volatile solids destruction and a decrease in gas production[1,2]. The magnitude of the impact on digestion performance depends on the amount of metal salt added and whether alum or ferric chloride is used. Higher dosages of metal salts reduce volatile solids destruction and gas production. Aluminum hydroxide can agglomerate on organic particles at high alum dosages. Coating of the surface of the organic particles could be a reason for the reduction in biodegradability[3].

The addition of metal salts to the wastewater will increase the solids loading to the digesters. Sludge concentrations can also be reduced, resulting in an increase in sludge volume which reduces the hydraulic residence time in digesters. The effect will depend on the metal salt dosage. Process design criteria for the digesters should be reevaluated when considering chemical phosphorus removal to make sure adequate capacity is available.

5.5.5 Effluent Phosphorus Limitation

Effluent phosphorus limitations and metal salt dosages are site-specific and can determine the impacts on sludge handling systems. Lower effluent phosphorus limitations require increased chemical dosage and can increase the impact on sludge handling facilities. High metal salt dosage and low wastewater alkalinity can reduce the effectiveness of anaerobic sludge digestion and require the addition of supplemental alkalinity.

5.5.6 Ultimate Disposal

The presence of iron or aluminum salt precipitates in a wastewater sludge will generally not affect the viability of any current ultimate sludge disposal practices. Numerous examples exist where sludges from full-scale wastewater treatment plants practicing chemical phosphorus removal are land applied. Sludges containing iron and aluminum precipitates can be safely used in agriculture as long as proper agricultural practices are followed.

Chemical addition may increase the heavy metal removal efficiency of the treatment plant, resulting in increased quantities of heavy metals in the sludge. The effect will be minimal in most cases since most plants efficiently remove influent heavy metals. However, the effect should be monitored and adjustments made in land application programs, as appropriate.

Chemical phosphorus removal sludges have been successfully incinerated. The quantity of residual ash certainly is increased, but the sludge can still be incinerated. The corrosivity of the sludge is increased when iron salts are present.

Landfilling can also be practiced, although the dry solids content of the dewatered sludge may decline. The major impact of chemical addition will be to increase the mass and volume of sludge which must be disposed of.

5.5.7 Cost

A preliminary indication of the impact of chemical phosphorus removal on solids handling costs at a WWTP is the relative increase in operating costs. The operator should determine the cost of current sludge handling systems as dollars per ton of dry solids processed. Estimated impacts of chemical addition on sludge thickening, dewatering, and digestion will negatively impact the unit cost. The adjusted unit cost can then be applied to the new, and larger, mass of sludge to be generated with metal salts addition. This gives the operator a value that is comparable among treatment options and between different WWTPs.

5.6 System Operation

5.6.1 Process Control

Automated controls for metal salt addition to wastewater for phosphorus are not well developed and the systems that have been installed do not have a good performance record. A fully automated system would adjust metal salt, polymer, and supplemental alkalinity dosage to maximize phosphorus removal and minimize cost with changing characteristics. Systems used to monitor orthophosphate concentration and adjust metal salt have not proven reliable. Adjustment of dosage based on influent flow can work if the characteristics of the wastewater are constant with time and with varying flow rates. In many systems, the wastewater characteristics are constantly changing.

A common process control system is to set the dose of the metal salt, polymer, and supplemental alkalinity based on observed performance. The dosages are typically set higher than needed to account for fluctuations in phosphorus loading without violation of permit limits at peak conditions. At low effluent phosphorus limits (less than 0.5 mg P/L) violation of the permit on a single day can raise the monthly average above allowable levels. At low limits, metal salt addition controls must be reliable and closely monitored.

5.6.2 Operating Costs

Operating and maintenance costs of chemical phosphorus removal systems are typically the largest component of the total present worth for the system. These costs should be determined during evaluation of the process and should be monitored closely during operation.

Operating costs include chemicals added to the wastewater, power, labor, maintenance, and costs of increased sludge handling. Table 5-7 lists some of the operating costs associated with chemical phosphorus control facilities. The values listed in Table 5-7 are generalized; values for a specific location can be quite different. Local suppliers should be contacted to obtain site specific values for detailed evaluation.

129

Table 5-7. Chemical phosphorus removal operating costs.

Item	Unit Cost
Polymer	$2.00/LB
Ferric Chloride (30%)	$0.95/GAL
Alum (Dry)	$0.13/LB
Alum (Liquid, 4.37% by wt)	$0.90/GAL
Sodium Hydroxide (50%)	$1.80/GAL
Electric Power	$0.04/KW
Operations Labor	$25/HR
Maintenance	2 percent of initial capital cost/yr

Operating costs also will increase due to additional quantities of solids recycled in the liquid treatment processes and increased mass of solids for disposal.

5.7 Full-Scale Experience

5.7.1 General

Chemical phosphorus removal systems are widely used throughout the U.S. and worldwide, and they have demonstrated the capability to reliably meet a variety of effluent discharge standards. Metal salts addition to primary and/or secondary treatment systems is widely practiced to meet a 1 mg P/L monthly average phosphorus standard as required for discharge to tributaries to the Great Lakes and elsewhere. Reliable compliance with the 1 mg P/L standard is achieved at many full-scale wastewater treatment plants. Metal salts addition followed by effluent filtration has been used successfully to meet monthly average effluent discharge standards of approximately 0.2 mg P/L. Discharge limits below this typically require the use of high lime treatment.

5.7.2 Case Studies

Because of the widespread, successful usage of chemical phosphorus removal systems, a great number of case histories are available. Summaries of operating and performance data from a wide variety of full-scale wastewater treatment plants can be found in the Water Pollution Control Federation Manual of Practice entitled **Nutrient Control**(7). Additional data are presented in two recent U.S. EPA manuals on phosphorus removal(1,4). Some of these data were previously presented in Chapter 4. These data bases provide extensive information which may assist the reader in evaluating chemical systems as an option for phosphorus removal.

Since a great deal of information is already available in the literature, an extensive description of case histories is not needed here. Consequently, four case histories were selected which represent a variety of operating and performance conditions. These case histories illustrate many, but not all, of the design and operating principles previously presented in this chapter. Three of the case histories presented involve the use of metal salts to meet monthly average effluent discharge standards ranging from 0.2 to 1.0 mg P/L, while one uses lime to meet a weekly average effluent discharge standard of 0.1 mg P/L. All three case histories of metal salt use involve the use of iron salts. This does not reflect a preference for iron salts over aluminum salts but rather a desire to illustrate the effects of other factors on overall system operation. Case histories using the same metal salt were selected so that this factor would be common between them.

The case histories presented are:

Type	Plant
Iron Salts Addition to Activated Sludge	Jones Island WWTP, Milwaukee, Wisconsin
Iron Salts Addition to Primary Clarifiers	South Shore WWTP, Milwaukee, Wisconsin
Multi-point Iron Salts Addition, Including Tertiary	Lower Potomac WPCP, Fairfax County, Virginia
Tertiary Lime Treatment	Upper Occoquan Sewage Authority, Virginia

5.7.2.1 Jones Island WWTP, Milwaukee, Wisconsin

Facility Description. The Jones Island Wastewater Treatment Plant (WWTP) is a 300 MGD peak flow secondary treatment plant with metal salts addition for phosphorus removal. As illustrated in Figure 5-4, influent wastewater receives preliminary treatment (bar screens and grit removal) and fine screening (3/32 inch slot width) before it is split between two parallel plug flow air activated sludge plants. The East Plant receives approximately 60 percent of the flow, while the West Plant receives the remaining 40 percent. Waste pickle liquor, a ferrous sulfate solution, is added to the East Plant influent. The iron is oxidized from the ferrous to ferric state in the aeration basins where it also precipitates phosphorus. Waste activated sludge from the East Plant is transferred to the West Plant; excess iron in the waste sludge serves as an iron source for phosphorus removal in the West Plant. Vacuum filter filtrate, which contains a sizeable quantity of residual iron from ferric chloride conditioning of the sludge, is also returned to the West Plant. Effluent from both the East Plant and West Plant is disinfected using chlorine and dechlorinated prior to discharge to Lake Michigan.

Waste sludge from the West Plant is gravity thickened and then dewatered by vacuum filters. The dewatered sludge is dried in rotary dryers and then packaged and marketed as a soil conditioner/fertilizer under the name Milorganite. The guaranteed analysis of Milorganite is 6 percent nitrogen and 1 percent phosphorus, minimum.

Effluent Limits. The Jones Island WWTP is a secondary treatment plant discharging to Lake Michigan. Its monthly average discharge standards are 30 mg $TBOD_5$/L, 30 mg TSS/L, and 1 mg TP/L.

Wastewater Characteristics. The Jones Island WWTP treats municipal wastewater, as well as most of the wastewater from heavy industry in the City of Milwaukee. Average wastewater characteristics for 1985-1986 are:

Influent Flow	138 MGD	
Influent $TBOD_5$	260 mg/L	(300,700 lb/day)
Influent TSS	210 mg/L	(244,300 lb/day)
Influent TP	5 mg/L	(6,000 lb/day)

An 0.5 percent phosphorus limit in laundry detergents is in effect in Wisconsin.

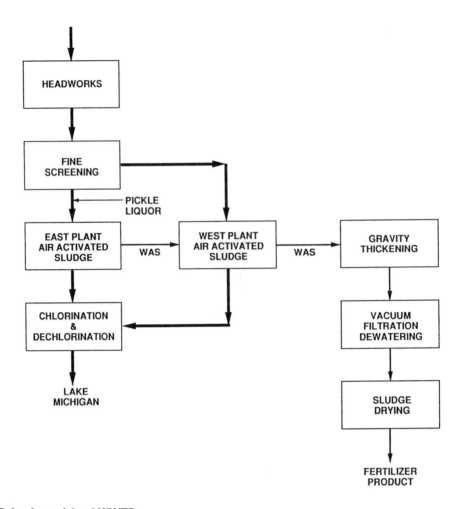

Figure 5-4. Jones Island WWTP.

Operating Results. Since implementation of solids handling system improvements in 1979, the plant has been almost continuously in compliance with its monthly average discharge standards. Figure 5-5 presents Jones Island influent and effluent phosphorus data for 1986. Also presented is the waste pickle liquor dose for each month, expressed as mg/L of iron. Jones Island receives waste pickle liquor free of charge from a local manufacturer. A probability plot of effluent total phosphorus concentrations from this plant is presented in Figure 5-10.

Summary. Jones Island is a highly successful example of metal salts addition to an activated sludge plant for phosphorus removal. Influent phosphorus concentrations are relatively low, due in part to the phosphorus ban for laundry detergents and to the presence of a relatively high proportion of high organic, low phosphorus strength industrial wastewater. Waste pickle liquor is delivered free of charge and is used to achieve effective phosphorus reduction through the facility. The addition of pickle liquor results in additional quantities of sludge to be disposed. However, sufficient capacity is available. Reliable performance is achieved by the system.

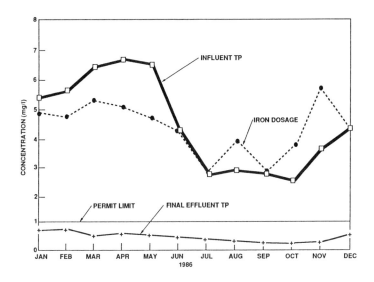

Figure 5-5. Jones Island WWTP phosphorus removal performance.

5.7.2.2 South Shore WWTP, Milwaukee, Wisconsin

Facility Description. The South Shore WWTP is a 200 MGD peak flow primary and secondary treatment plant with metal salts addition for phosphorus removal. As illustrated in Figure 5-6, influent wastewater receives preliminary treatment (bar screens and grit removal) prior to primary clarification and biological treatment in a plug flow air activated sludge system. Activated sludge effluent is disinfected with chlorine and dechlorinated prior to discharge to Lake Michigan.

Waste pickle liquor is added to the primary clarifier influent for phosphorus removal. Waste pickle liquor as received at the plant contains iron primarily in the ferrous ($+2$) state. Prior to its addition to the primary influent, chlorine is used to oxidize the pickle liquor. This converts the ferrous ($+2$) iron to ferric ($+3$) iron. Previous experience at South Shore indicated that this was necessary to allow effective utilization of the applied iron dose in the primary clarifiers.

Waste activated sludge is thickened in dissolved air flotation units and then anaerobically digested with the primary sludge. Digested sludge is lagooned prior to agricultural reuse. Lagoon supernatant is returned to the plant headworks.

Effluent Limits. Effluent limits for South Shore are identical to those for Jones Island. The monthly average discharge limits are 30 mg $TBOD_5$/L, 30 mg TSS/L, and 1 mg TP/L.

Wastewater Characteristics. The wastewater received at South Shore is largely domestic and commercial in nature. Average wastewater characteristics for 1985-1986 are:

Influent Flow	100 MGD	
Influent $TBOD_5$	138 mg/L	(115,300 lb/day)
Influent TSS	169 mg/L	(141,200 lb/day)
Influent TP	5 mg/L	(4,000 lb/day)

This plant is also affected by the ban on phosphorus in laundry detergents.

133

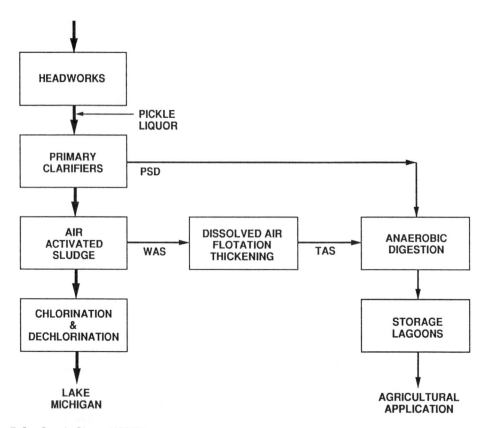

Figure 5-6. South Shore WWTP.

Operating Results. Prior to 1983, pickle liquor was added to the South Shore secondary treatment system just like at Jones Island. Pilot studies demonstrated that iron doses could be reduced significantly through the use of oxidized pickle liquor. Addition to the primary clarifiers offered two advantages: (1) an increase in the quantity of more easily handled primary sludge and a decrease in the quantity of more difficult to handle waste activated sludge and (2) improved performance since the phosphorus content of the activated sludge mixed liquor was reduced (i.e., most of the phosphorus was precipitated and removed in the primary sludge). No impact of iron addition on anaerobic digester gas production or on the quality of the lagoon supernatant was noted.

Figure 5-7 presents operating and performance data for 1986 at South Shore. Iron doses averaged approximately 1 mg/L as iron per mg P/L in the influent wastewater. At these doses the effluent total phosphorus concentration was reliably below the monthly average discharge limit of 1 mg P/L. A probability plot of effluent total phosphorus concentrations from this plant is presented in Figure 5-10.

Summary. The South Shore case history, when compared to Jones Island, illustrates the advantages of addition of metal salts to the primary treatment system rather than to the secondary treatment system. Both plants are owned and operated by the same agency, the Milwaukee Metropolitan Sewerage District, and both use the same chemical for phosphorus removal (waste pickle liquor). Primary clarifiers are not available at Jones Island, so pickle liquor is added to the secondary process. At South Shore, chemical is added to the primary clarifiers because of reduced impacts on the solids handling system and because addition at this point results in superior performance at South Shore.

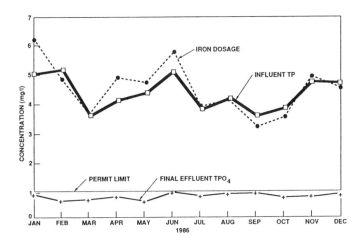

Figure 5-7. South Shore WWTP phosphorus removal performance.

5.7.2.3 Lower Potomac WPCP, Fairfax County, Virginia

Facility Description. The Lower Potomac Water Pollution Control Plant (WPCP) is a 36 MGD advanced wastewater treatment (AWT) plant. Figure 5-8 is a process schematic for the treatment plant. The plant incorporates preliminary treatment (bar screens), primary clarification, secondary treatment using plug flow air activated sludge, flow equalization, tertiary chemical addition/clarification, filtration, chlorination, and dechlorination. Primary sludge is degritted and gravity thickened, and waste activated sludge is thickened in dissolved air flotation units. Thickened primary and waste activated sludge is blended, dewatered in vacuum filters, and incinerated. Chemical sludge is gravity thickened, dewatered in centrifuges, and landfilled.

Lower Potomac was originally designed to utilize high-lime tertiary treatment with two-stage recarbonation for phosphorus removal. Due to difficulties encountered, in 1980 the AWT facilities were converted to remove phosphorus using ferric chloride. As illustrated in Figure 5-8, a multi-point addition system is used for the iron salts. Ferrous sulfate is added to the influent wastewater, and ferric chloride and polymer are added at two points in the process flow stream: (1) to the activated sludge mixed liquor as it flows to the secondary clarifiers and (2) to the AWT influent. Ferrous sulfate is added to the influent wastewater because of the superior sludge handling characteristics of the resulting primary sludge.

Effluent Limits. The Lower Potomac WPCP discharges to the Potomac River. Its discharge permit includes monthly average limitations of 8 mg $TBOD_5$/L, 8 mg TSS/L, and 0.2 mg TP/L.

Wastewater Characteristics. Wastewater treated at Lower Potomac is primarily domestic and commercial in nature. Average wastewater characteristics for fiscal year 1987 are:

Parameter	Average
Flow, MGD	33
$TBOD_5$, mg/L	177
TSS, mg/L	215
TP, mg/L	7

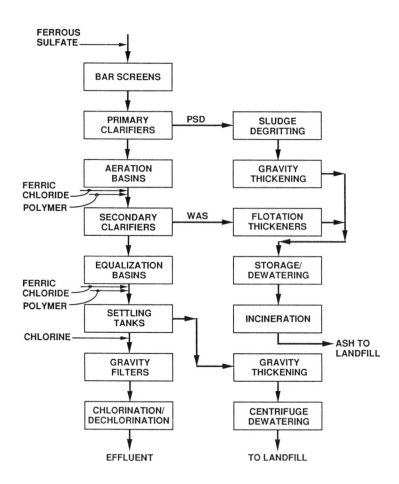

Figure 5-8. Lower Potomac WWTP.

Operating Results. The average iron dose at Lower Potomac during fiscal year 1987 was 17 mg/L as iron, which corresponded to an iron dose of 2.5 mg/L of iron per mg TP/L removed. The doses were 6 mg/L of iron to the primary treatment system, 5 mg/L of iron to the secondary treatment system, and 6 mg/L of iron to AWT. The chemical cost was reported to be $0.54 per lb of TP removed. However, the total cost, including chemicals, labor, utilities, sludge disposal, and administration, was reported to be $4.00 per lb of TP removed.

Effluent quality for fiscal year 1987 averaged 7 mg $TBOD_5$/L, 0.8 mg TSS/L, and 0.12 mg TP/L. The effluent TP limit of 0.2 mg P/L was reliably met. A probability plot of effluent total phosphorus concentrations from this plant is presented in Figure 5-10.

Summary. The lower Potomac case history illustrates the use of multi-point metal salts addition and effluent filtration to reliably produce an effluent with TP less than 0.2 mg P/L. Experience at other facilities (such as Blue Plains in the District of Columbia) indicates that tertiary clarification is not necessary, but operators at Lower Potomac indicate that it is beneficial. Multi-point chemical addition provides the opportunity to optimize chemical addition to minimize chemical costs and sludge handling impacts. This case history also indicates that the costs for sludge handling arising out of chemical addition may be far greater than the costs for chemicals alone.

136

5.7.2.4 Upper Occoquan Sewage Authority, Virginia

Facility Description. The Upper Occoquan Sewage Authority (UOSA) Regional Water Reclamation Plant (RWRP) is a 15 MGD advanced wastewater treatment (AWT) facility which treats domestic and commercial wastewater to extremely high levels for discharge to the Occoquan Reservoir, the principal water supply reservoir in Northern Virginia.

As indicated in Figure 5-9, the plant consists of preliminary treatment (coarse screening, comminution, and grit removal), primary clarification, secondary treatment using complete mix air activated sludge, high lime treatment with two-stage recarbonation, flow equalization, filtration, activated carbon, post filtration, and chlorination. An ion exchange system is available for total nitrogen removal. However, due to the current discharge standards and the high cost associated with the ion exchange system, ammonia removal by nitrification in the air activated sludge system is the only form of nitrogen control practiced.

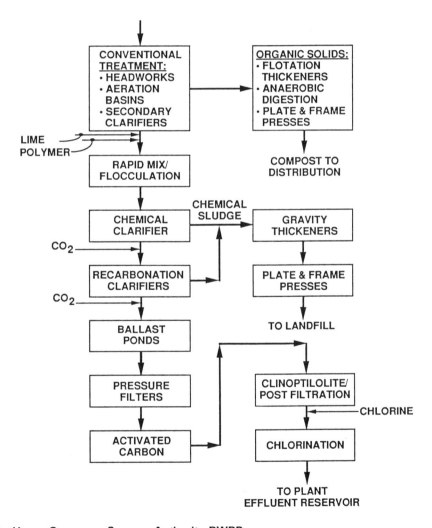

Figure 5-9. Upper Occoquan Sewage Authority RWRP.

Waste activated sludge is thickened in dissolved air flotation units, mixed with the primary sludge, and stabilized in anaerobic digesters. Digested sludge is chemically conditioned using lime and ferric chloride, dewatered in plate and frame presses, and composted. Chemical sludge is gravity thickened, dewatered in plate and frame presses, and landfilled.

Effluent Limits. Effluent limits for the UOSA RWRP are specified on a weekly rather than on a monthly basis. Weekly limits are 1.0 mg $TBOD_5$/L, 1.0 mg TSS/L, 0.1 mg TP/L, 1.0 mg TKN-N/L, 10 mg COD/L, 0.4 JTU for turbidity, 0.1 mg MBAS/L, and 2 fecal coliforms per 100 mL.

Wastewater Characteristics. The wastewater treated at UOSA is primarily domestic and commercial in nature. During 1984 the wastewater characteristics were as follows:

Influent Flow, MGD	10.2
Influent $TBOD_5$, mg/L	200
Influent TSS, mg/L	170
Influent TP, mg P/L	9
Influent TKN, mg N/L	34

Operating Results. The UOSA RWRP has demonstrated an outstanding record of compliance with its discharge permit. For 1984 (a year of typical performance) the median effluent quality was as follows:

Effluent $TBOD_5$, mg/L	0.5
Effluent COD, mg/L	6.8
Effluent TSS, mg/L	0.1
Effluent TP, mg P/L	0.03
Effluent TKN, mg N/L	0.4

The 0.1 mg TP/L weekly average discharge limit was met each week, and daily average values were less than 0.1 mg TP/L more than 95 percent of the time. A probability plot of effluent total phosphorus concentrations from this plant is presented in Figure 5-10.

Nitrification in the secondary treatment process reduces the alkalinity of the secondary effluent to approximately 70 to 80 mg/L as $CaCO_3$. Consequently, the typical lime dose is 190 to 200 mg/L as CaO. Operating costs for the chemical treatment system, including solids handling were reported to be $235 per million gallons of wastewater treated.

Summary. An extremely high level of treatment is provided at UOSA to allow reuse of the treated effluent. As such, treatment goes beyond just phosphorus removal. However, the results obtained at UOSA demonstrate the high level of performance, and extremely high level of reliability, which can be achieved with this level of technology. Average effluent TP concentrations are extremely low and effluent quality meets the permit limit of 0.1 mg TP/L the vast majority of the time. As might be expected, costs to provide this level of treatment capability and reliability are high.

5.7.2.5 Conclusion

The four case histories presented above illustrate the treatment capabilities and reliability of chemical phosphorus removal systems. A comparison is provided in Figure 5-10.

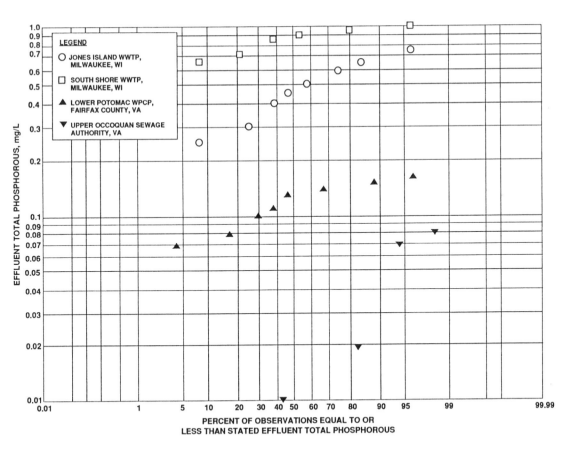

Figure 5-10. Probability plot of monthly average effluent total phosphorus concentrations.

All four facilities reliably meet their effluent phosphorus discharge standards. The Jones Island and South Shore case histories illustrate the use of metal salts addition to primary and secondary treatment systems to meet a 1 mg TP/L monthly limit. The Lower Potomac case history illustrates the use of multi-point metal salts addition and effluent filtration to meet a very stringent effluent total phosphorus limit (0.2 mg P/L), while the UOSA case history illustrates the performance capability of high lime treatment followed by extensive filtration.

5.8 References

1. U. S. Environmental Protection Agency. **Design Manual: Phosphorus Removal**. EPA/625/1-87/001, 1987.

2. Gossett, J. M., et al. Anaerobic digestion of sludge from chemical treatment. **Jour. Water Pollut. Control Fed.**, **50**, 533, 1978.

3. Dentel, S. K. and J. M. Gossett. Coagulation of organic suspensions with aluminum salts. **Jour. Water Pollut. Control Fed.**, **59**, 101, 1987.

4. U. S. Environmental Protection Agency. **Handbook: Retrofitting POTWs for Phosphorus Removal in the Chesapeake Bay Drainage Basin.** EPA/625/6-87/016, 1987.

5. American Water Works Association. **Simplified Procedures for Water Examination.** AWWA Manual M-12.

6. E. I. DuPont De Nemours & Co. Ferric chloride product data publication.

Chapter 6

Principles of Biological Phosphorus Removal

6.1 Introduction

A typical phosphorus content of microbial solids is 1.5 to 2 percent on a dry weight basis. Wasting of excess solids with this phosphorus content from a municipal activated sludge plant may result in 10-30 percent phosphorus removal. For example, assuming a primary effluent BOD_5 concentration of 120 mg/L, a soluble phosphorus concentration of 8 mg P/L, and a waste solids yield of 0.60 g VSS/g BOD_5, a removal of 1-1.5 mg P/L of soluble phosphorus would result, for a 12-19 percent removal efficiency.

Biological phosphorus removal involves design or operational modifications to conventional treatment systems that results in the growth of a biological population that has a much higher cellular phosphorus content. Such systems incorporate an anaerobic operating phase somewhere in the process, and the waste sludge overall phosphorus content is typically in the range of 3-6 percent. This diverts more phosphorus to the waste solids and yields lower effluent phosphorus concentrations.

The evolution of the design and application of biological phosphorus removal systems is unique in the field of Sanitary Engineering. The phenomenon was unknowingly observed in full-scale plants in the early 1960s, but only after research in the early 1970s identified the necessary operating conditions did intentional process designs occur for full-scale facilities.

The objective of this chapter is to describe the historical background of biological phosphorus removal, the fundamental biological mechanism responsible for biological phosphorus removal, the system designs, and critical process and design considerations that affect the performance of biological phosphorus removal systems.

6.2 Historical Background

The historical development of biological phosphorus removal systems involved a sequence of: 1) observations on sludges and full-scale plants that had significant phosphorus removal capacities, 2) the recognition of the need for an anaerobic contact zone for sludge prior to an aerobic zone, 3) the need to exclude anoxic or aerobic electron acceptors from the anaerobic zone, and 4) the role and need of simple substrates in the anaerobic zone.

As early as 1955, Greenburg et al.(1) proposed that activated sludge could take up phosphorus at levels beyond the normally accepted microbial growth requirements. Srinath(2) and Alarcon(3) were the first researchers to report the occurrence of biological phosphorus removal from wastewater treatment plant sludges. Both observed rapid phosphorus uptake when sludge samples taken from a local plug flow activated sludge plant were mixed and aerated with raw wastewater. However, they could not explain this phenomenon.

Levin and Shapiro(4) coined the term "Luxury Uptake" of phosphorus after they observed enhanced biological phosphorus removal using activated sludge from the District of Columbia activated sludge plant. Over 80 percent phosphorus removal was reported after vigorous aeration of the sludge. When they added 2-4 di-nitrophenol to the reactors, the phosphorus uptake during aeration was inhibited to suggest that the high phosphorus removal was of biological origin. They also reported observing volutin granules in the cells which are compounds that are known to contain polyphosphates. When they held the sludge under anaerobic conditions or acidified it, phosphorus release took place. No explanation was offered to explain the phosphorus release, but the first commercial biological phosphorus removal process was developed from this work; namely the Phostrip process.

Besides Levin and Shapiro's work, high levels of phosphorus removal was also reported at a number of full-scale facilities, which included the Rillings Road plant in San Antonio(5), the Hyperion plant in Los Angeles(6), and the Back River plant in Baltimore(7). All of these plants were conventional, long, narrow, plug flow tank designs with elevated dissolved oxygen (DO) concentrations occurring towards the end of the aeration tanks.

The San Antonio plant showed 88 percent phosphorus removal with 4.3-7.3 percent phosphorus in the sludge on a dry weight basis. Effluent phosphorus concentrations at the Hyperion plant ranged from 0.5-1.8 mg P/L. Rapid phosphorus uptake occurred in the first half of the tank when the DO was elevated. Phosphorus release was noted near the tank inlet which followed a tank that was used to distribute return sludge and primary effluent to the aeration trains. A significant detention time occurred in this tank and anaerobic conditions were very likely. A similar activated sludge plant in the area did not have the contact tank design for feed distribution, and it did not show any unusual phosphorus uptake as at Hyperion.

The Baltimore plant reported 88 percent phosphorus removal and 2-5 percent phosphorus in the waste sludge. The phosphorus uptake occurred in the latter part of the tank and was associated with increased DO levels. Milbury(7) reported that phosphorus release also was occurring at the front end of the tank.

During this period of reports of high phosphorus removal there were varying opinions as to whether it was due to a biological mechanism or chemical precipitation. There was very little data or experience to explain the biological mechanism, but the tank designs and pH changes favored the chemical precipitation theory(8). Due to the consistent observations of higher pH and high aeration rates with possible carbon dioxide stripping at the end of the plug flow tanks cited above, formation of a phosphorus precipitate as calcium hydroxy apatite seemed plausible.

Empirical design guidelines were proposed by Vacker et al.(5) and Milbury(7) that provided an important basis for future studies. They recognized that the following operating conditions favored biological phosphorus removal: 1) a plug flow tank with wastewater added only at the inlet end, and 2) reversed tapered aeration with a sufficient DO concentration of greater than 2 mg/L at the downstream end of the tank and avoidance of nitrification. Though not specifically stated, these guidelines showed the importance of cycling sludge under alternating anaerobic/aerobic conditions and the need for substrate addition to the anaerobic zone.

During the early 1970s the development of the Phostrip process continued and the key operating condition for biological phosphorus removal was reported by Barnard(9). Barnard reported that efficient phosphorus removal could occur biologically in a system, where the sludge was subjected to an anaerobic state of sufficient intensity to release phosphorus, followed by an aerobic operating phase. This also provided explanation for the performance of the full-scale, plug flow plants that experienced high levels of phosphorus removal. Lack of a sufficient oxygen transfer rate resulted in anaerobic conditions at the inlet zones of the plug flow tanks and an associated release of phosphorus. Phosphorus uptake occurred later in the tanks in zones of elevated DO.

In a later paper, Barnard(10) proposed the use of a separate anaerobic basin ahead of the aerated activated sludge basin and termed the process the Phoredox process (Figure 6-1). Phoredox was derived from phosphorus and redox potential to signify the lower reduced conditions required in the anaerobic zone. Barnard also noted that the presence of nitrates in the anaerobic zone had an adverse affect on the biological phosphorus removal efficiency. Experiments in pilot plant and full-scale facilities confirmed the negative impact of nitrates on biological phosphorus removal(11,12).

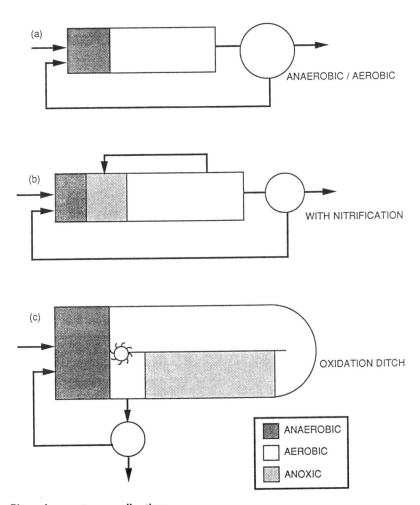

Figure 6-1. Phoredox system applications.

Following Barnard's pilot plant work, full-scale facilities were modified at Johannesburg, South Africa to investigate the feasibility of biological phosphorus removal. At the Alexander plant, surface aerators in the inlet zone of an activated sludge basin were turned off to create an anaerobic-aerobic treatment sequence(13). Overall nitrogen and phosphorus removal efficiencies of 85 and 46 percent, respectively, were reported. At the Olifantsvlei plant various combinations of surface aerators were turned off in the four stage system and an effluent soluble phosphorus concentration of 0.9 mg/L was achieved(14). Based on this work, a 39 MGD nutrient removal facility using the modified Bardenpho process was designed and started in 1978(15).

In the late 1970s biological phosphorus removal facilities using anaerobic aerobic zones were started up in the United States at Palmetto, Florida and Largo, Florida(16,17).

Nicholls and Osborn(12) provided direction for further understanding and modification of the anaerobic-aerobic biological phosphorus removal system. They proposed a biochemical model involving carbon storage products, such as polyhydroxybutyrate (PHB), and polyphosphates to explain biological phosphorus removal. Under an anaerobic "stressed" condition, simple substrates would be stored as PHB, and this was somehow linked to phosphorus release. Under aerobic conditions, the PHB would be degraded to produce energy to be made available for polyphosphate storage. With the recognition that simple substrates formed from fermentation were important to the process, they recommended feeding supernatant from anaerobic digestion of primary sludge to the anaerobic zone to improve phosphorus removal performance.

6.3 Biological Phosphorus Removal Mechanism

Initial explanations for the mechanism of biological phosphorus removal referred to the anaerobic zone as providing a "stressed" condition that resulted in phosphorus release which in turn was followed by an aeration zone where enhanced phosphorus uptake occurred. As more information was gained, a generally accepted mechanistic model evolved that includes fundamental biochemical considerations. Understanding the removal mechanism leads to more rational and improved designs and a better appreciation of conditions that affect the process performance.

The biological phosphorus removal mechanism is based on the following key facts:

1)   Bacteria are capable of storing excess amounts of phosphorus as polyphosphates.

2)   These bacteria are capable of removing simple fermentation substrates produced in the anaerobic zone and assimilating them into storage products within their cells. This process involves the release of phosphorus.

3)   In the aerobic zone, energy is produced by the oxidation of storage products and polyphosphate storage in the cell increases.

One term used to describe the anaerobic zone is that it is a "biological selector" for phosphorus-storing microorganisms. This zone provides a competitive advantage for the phosphorus-storing microorganisms, since they can take up substrate in this zone before other, non-phosphorus-storing bacteria can. Thus, this zone allows the development or selection of a large population of phosphorus-storing organisms in the system which take up significant levels of phosphorus and are removed from the system via the waste sludge.

144

An important benefit from this population selection for biological phosphorus removing organisms is the resultant prevention of the proliferation of filamentous bacteria that cause poor sludge settling characteristics. Thus, by employing the biological phosphorus removal anaerobic/aerobic treatment sequence it is possible to develop a mixed liquor with relatively low sludge volume index (SVI) values.

6.3.1 Fate of Substrate in the Anaerobic Zone

Figure 6-2 shows a typical profile of soluble BOD (SBOD) and orthophosphorus (Pi) in the anaerobic and aerobic zones of a Phoredox system. The SBOD concentration decreases in the anaerobic zone even though there is no aerobic or anoxic electron acceptor present. While the SBOD concentration decreases the soluble Pi concentration also increases in the anaerobic zone and is taken up later to low concentrations in the aerobic zone. As an example of this behavior, Hong et al.(18) reported a SBOD concentration decrease from 45 to 15 mg/L and a soluble Pi increase from 6 to 24 mg P/L in the anaerobic zone of a biological phosphorus removal system.

Nicholls and Osborne(12) proposed that short chain fatty acids, such as acetate, are produced in the anaerobic zone as a result of fermentation reactions. Observations of an increase in PHB storage products during the anaerobic contact period helped to confirm the role of fatty acids, as well as explain the disappearance of SBOD. Increased PHB concentrations were identified by Timmerman(19) as well as Nicholls and Osborne(12). Deinema(20) also observed PHB in a strain of phosphorus removing **Acinetobacter**. Buchan(21) reported that during the PHB accumulation in the cell in the anaerobic zone, polyphosphate granules decreased in size or disappeared.

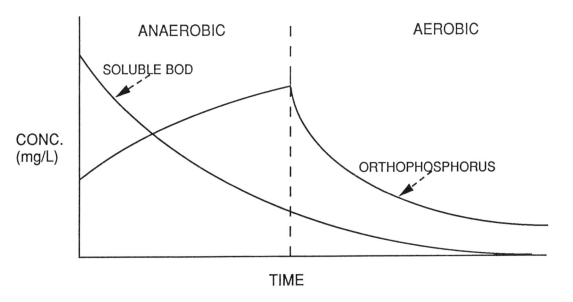

Figure 6-2. Fate of soluble BOD and phosphorus.

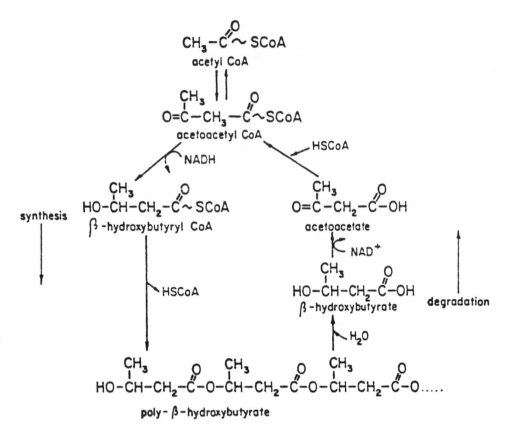

Figure 6-3. Poly-ß-hydroxybutyrate metabolic pathways(22).

PHB synthesis for bacterial cells is shown in Figure 6-3(22). PHB is formed from acetoacetate serving as an electron acceptor allowing the reoxidation of NADH into NAD without using the oxidative electron transport chain. The conversion of acetate to acetyl COA requires energy from within the cell. During aerobic conditions PHB is oxidized to acetyl COA which can enter the TCA cycle. This oxidation is accomplished by the bacterial cell provided that other degradable substrates are not available.

Other polyhydroxyalkanoates have been found in biological phosphorus-storing microorganisms. A common one has been identified as polyhydroxyvalerate (PHV) by Comeau(22). PHV is formed from acetate and propionate entering the cell under anaerobic conditions. The assimilated carbon storage compounds may be able to accumulate to a point where they represent up to 50 percent of the dry weight of the biological cell. PHB storage can be identified in a cell by staining, by spectrophotometric techniques and by extraction and gas chromatography analysis.

The observations on PHB and PHV storage suggests that the preferred fermentation products for phosphorus storing organisms are acetate and propionate. Table 6-1 shows that investigators have found a molar ratio relationship between acetate fed to biological phosphorus-removing cultures and the soluble Pi released in the anaerobic zone. Through batch experiments other investigators have shown that acetate and propionate are preferred substrates to stimulate phosphorus release in the anaerobic zone(27,28,29).

146

Table 6-1. Acetate affects phosphorus release in anaerobic zone.

Moles Acetate Added/Mole P Release	Reference
0.7	22
1.0	23
0.7	24
0.6	25
1.0	26

Experiments by Gerber et al.(30) illustrated the role of simple volatile fatty acids (VFAs) and nitrate in the anaerobic zone of biological phosphorus removal systems. Nitrates with different short chain carbohydrates were fed to a number of batch reactors containing biological phosphorus-removing sludge. Compounds used included acetic acid, propionic acid, butyric acid, lactic acid, formic acid, citric acid, succinic acid, glucose, ethanol, and methanol. The organic substrate, nitrate and soluble Pi concentrations were measured with time under anaerobic conditions. Phosphorus release in the presence of nitrate only occurred for reactors that contained acetic, propionic, or formic acids. Phosphorus release did not occur with the other compounds until the nitrates were reduced. Since nitrate would interfere with fermentation, this showed that the other compounds had to be converted to the preferred substrates before phosphorus release could occur.

Buchan(21) showed that concurrent with substrate uptake and phosphorus release in the anaerobic zone, volutin granules dispersed into smaller granules or disappeared. As mentioned, volutin granules were observed during Levin and Shapiro's(4) early work on phosphorus removal. Such granules are known to contain lipids, protein, RNA and magnesium in addition to polyphosphates(31). The granules are visible under the light microscope and can be identified by staining with either toluidine dye, which results in a reddish purple color, or with a methylene blue technique which results in a dark purple color. A high electron beam directed on the microorganisms will also volatilize the polyphosphates leaving holes in the volutin granules in the cell. As will be explained in a summary of the biological phosphorus removal model, the polyphosphates play a major role of energy storage and release to facilitate the substrate storage abilities of phosphorus-storing organisms under anaerobic conditions.

Release and uptake of metal ions has also been observed with the release and uptake of soluble Pi(32). The most common cations released with soluble Pi released are magnesium and potassium, as well as a small amount of calcium. Typical molar ratios of cation to phosphorus release are about 0.28, 0.26 and 0.09 for magnesium, potassium and calcium, respectively. On a charge basis these cations account for most of the negative charge associated with the release of soluble orthophosphorus.

6.3.2 Phosphorus Storing Microorganisms

Microbiological literature indicates that a number of microorganisms are capable of storing excess amounts of phosphorus in their cells. Fuhs and Chen(33) were one of the first investigators to isolate phosphorus-storing organisms. This was done with sludges from the Baltimore Back River and Seneca Falls plants, which were both exhibiting high levels of phosphorus removal. They identified the organism associated with phosphorus removal as **Acinetobacter**. These bacteria are short, plump, gram-negative rods with a size of 1-1.5 μm. They appear in pairs, short chains or clusters. These bacteria are known to prefer simple substrates as would be produced in fermentation reactions in anaerobic zones of biological phosphorus removal systems.

Other bacteria commonly found in biological phosphorus removal systems are species of **Pseudomonas** and **Aeromonas**. **Pseudomonas** appear to be responsible for biological phosphorus uptake, while **Aeromonas** appear to be important for accomplishing fermentation and VFA production.

Since many biological phosphorus removal systems involve nitrification and denitrification, the ability for phosphorus-storing microorganisms to reduce nitrate is an important issue. Since they have the potential to remove a large portion of the influent BOD, denitrification rates in an anoxic zone following the anaerobic zone should be lower compared to possible rates without the anaerobic zone. However, many investigators have now observed phosphorus uptake in anoxic zones concurrent with nitrate reduction(33,34,35), but it is not known if the denitrification rate is equivalent to that possible if no anaerobic zone and substrate storage exist. It is generally believed that biological oxidation reactions are faster using readily available SBOD than intracellular storage products.

6.3.3 Summary of Biological Phosphorus Removal Mechanism

The biological phosphorus removal mechanism is summarized in Table 6-2. Acetate and other fermentation products are produced from fermentation reactions by normally occurring facultative organisms in the anaerobic zone. A generally accepted concept is that these fermentation products are derived from the soluble portion of the influent BOD and that there is not sufficient time for the hydrolysis and conversion of the influent particulate BOD. The fermentation products are preferred and readily assimilated and stored by the microorganisms capable of excess biological phosphorus removal. This assimilation and storage is aided by the energy made available from the hydrolysis of the stored polyphosphates during the anaerobic period. The stored polyphosphate provides energy for active transport of substrate and for formation of acetoacetate, which is converted to PHB.

Table 6-2. Biological phosphorus removal steps.

Anaerobic Zone

1. Fermentation: SBOD converted to VFAs by facultative
                                            organisms

2. Biological P storing organism obtains VFA: VFA transferred into cell
                                            Orthophosphorus release provides energy
                                            VFA converted to PHB/PHV

Aerobic Zone

1. Phosphorus uptake: PHB oxidized
                                            Energy captured in polyphosphate bonds
                                            Orthophosphorus removed from solution

2. New cells produced

System Phosphorus Removal

1. Excess sludge wasting: Phosphorus removal via wasted sludge

The fact that phosphorus-removing microorganisms can assimilate the fermentation products in the anaerobic phase means that they have a competitive advantage compared to other normally occurring microorganisms in activated sludge systems. Thus, the anaerobic phase results in selection and development of a population of phosphorus-storing microorganisms. Rensink et al.(36) pointed out that **Acinetobacter** are relatively slow growing bacteria and that they prefer simple carbohydrate substrates. Thus, without the anaerobic phase, they may not be present at significant levels in conventional activated sludge systems.

During the aerobic phase, the stored substrate products are depleted and soluble phosphorus is taken up, with excess amounts stored as polyphosphates. An increase in the population of phosphorus-storing bacteria is also expected as a result of substrate utilization.

The above mechanism indicates that the level of biological phosphorus removal achieved is directly related to the amount of substrate that can be fermented by normally occurring microorganisms in the anaerobic phase and subsequently assimilated and stored as fermentation products by phosphorus removing-microorganisms, also in the anaerobic phase.

6.4 Biological Phosphorus Removal Systems

This section introduces various flow sheets used for biological phosphorus removal. The design of the systems are presented in more detail in Chapter 7. The common feature in all of the systems is the use of an anaerobic zone for substrate uptake by phosphorus-storing bacteria. Many design modifications are a result of nitrification and denitrification considerations. The three common commercial biological phosphorus removal processes are shown in Figure 6-4.

The Phostrip process was first proposed by Levin in 1965(4). Pilot plant data were collected at a number of municipal plants from 1970 to 1973 and demonstrated high levels of phosphorus removal. In 1973, the Seneca Falls, New York activated sludge plant was converted to the Phostrip process and evaluated (37). The process combines both biological and chemical phosphorus removal and has been referred to as a sidestream process, since a portion of the return activated sludge flow is diverted for phosphorus stripping and subsequent precipitation with lime.

The Modified Bardenpho process, marketed by the Eimco Process Equipment Company in Salt Lake City, Utah, is both a nitrogen and a phosphorus removal system. As Figure 6-4 illustrates, the influent and return sludge are contacted in an anaerobic tank to promote fermentation reactions and phosphorus release prior to passing the mixed liquor through the four-stage Bardenpho system.

The original development of the four-stage Bardenpho process was to provide for more than 90 percent nitrogen removal without using an exogenous carbon source. In the first anoxic stage, nitrate-nitrogen contained in the internal recycle from the nitrification stage is reduced to nitrogen gas (denitrification) by metabolizing influent BOD using nitrate oxygen instead of DO. About 70 percent of the nitrate-nitrogen produced in the system is removed in the first anoxic stage. In the nitrification (first aerobic) stage, BOD removal, ammonium-nitrogen oxidation, and phosphorus uptake occurs. The second anoxic stage provides sufficient detention time for additional denitrification by mixed liquor endogenous respiration, again using nitrate oxygen instead of DO. The final aerobic stage provides a short period of mixed liquor aeration prior to clarification to minimize anaerobic conditions and phosphorus release in the secondary clarifier.

The A/O process is marketed in the United States by Air Products and Chemicals, Inc. in Allentown, Pennsylvania and is similar to the Phoredox concept described in Figure 6-1, except that the anaerobic and aerobic stages are divided into a number of equal size complete mix compartments. Typically, three compartments have been used for the anaerobic stage and four for the aerobic stage. The key features of the A/O process are its relatively short design SRT and high design organic loading rates.

The A/O process can also be used where nitrification and/or denitrification are required. The modified flow scheme incorporates an anoxic stage for denitrification between the anaerobic and aerobic stages and is called the A^2/O process. The anoxic stage is also divided into three equal-size, complete mix compartments. Mixed liquor is recycled from the end of the nitrification stage to feed nitrate-nitrogen into the anoxic stage for denitrification. Internal recycle flows of 100-300 percent have been used. Nitrate-nitrogen removals of 40-70 percent can be accomplished this way.

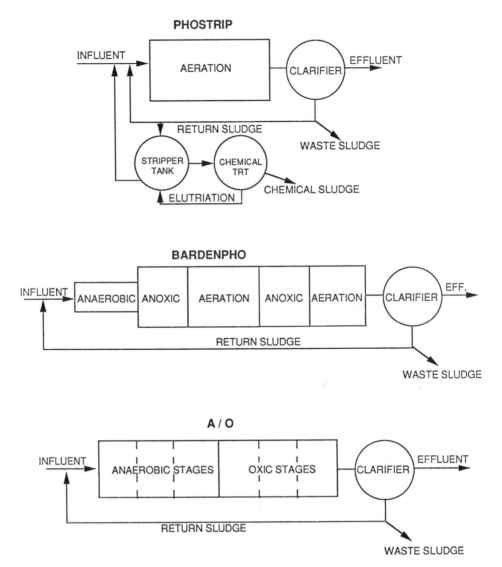

Figure 6-4. Biochemical and biological phosphorus removal systems.

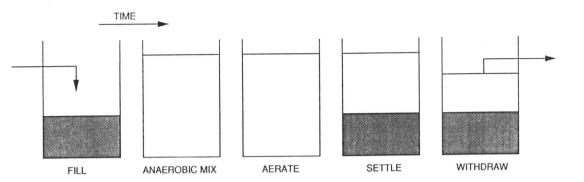

TIME

FILL   ANAEROBIC MIX   AERATE   SETTLE   WITHDRAW

Figure 6-5. Biological phosphorus removal using a sequencing batch reactor.

The use of sequencing batch reactor (SBR) systems for secondary treatment has gained increased popularity in the United States during the late 1970s and early 1980s. An evaluation of SBR treatment capabilities, design aspects, full-scale installations, and advantages has been documented for conventional activated sludge treatment applications(38). Though not a new treatment concept, with reported operations dating back to the early 1900s, the recent surge of interest has been related to new and improved hardware devices and to the successful EPA-funded, full-scale, 20-month demonstration and evaluation of a 1,330 m³/d (0.35 MGD) facility at Culver, Indiana(39).

Unique hardware for the SBR system consists of motorized or pneumatically-actuated valves, level sensors, automatic timers, microprocessor controllers, and effluent withdrawal decanters. The SBR treatment concept and operational flexibility makes it an obvious candidate for employing anaerobic-aerobic contacting for biological phosphorus removal. Biological phosphorus removal was demonstrated in the full-scale Culver, Indiana facility during June and July 1984(40).

A schematic of an SBR operation for biological phosphorus removal is shown in Figure 6-5. The SBR system is a fill-and-draw activated sludge system. A single tank provides for activated sludge aeration, settling, effluent withdrawal, and sludge recycle. The operation steps consist first of a fill period where flow is diverted to one of the SBR tanks while the other tank(s) operates in the reaction, settle, effluent withdrawal, or idle operation sequences. After the fill period the reactor contents are mixed, but not aerated, to provide the anaerobic fermentation period for phosphorus release and uptake of soluble fermentation products. The next step is the react or aeration period followed by a selected settling time when both aeration and mixing are stopped. The effluent is then withdrawn and, depending on the influent flow rate, a variable length idle time may occur.

Figure 6-6 shows a further modification of the Modified Bardenpho process. This modification was developed at the University of Capetown in South Africa(41) and has been termed the UCT process. As shown, the return activated sludge is directed to the anoxic stage instead of the anaerobic stage as in the Modified Bardenpho process. The basis for this development was previous work with biological phosphorus removal systems that indicated initial phosphorus removal efficiency could be negatively affected by nitrate-nitrogen entering the anaerobic stage. Nitrate will serve as an electron acceptor during the biological oxidation of BOD entering the anaerobic stage. This results in competition for the soluble, readily biodegradable BOD that would normally be converted to fermentation products for use by the biological phosphorus-removing bacteria in the anaerobic zone in the absence of nitrate-nitrogen. The relative ratio between the nitrate-nitrogen in the return sludge to the readily degradable soluble BOD in the influent to the anaerobic zone of a Phoredox or A/O

151

Process will determine if sufficient BOD will remain after denitrification reactions occur to produce a necessary level of fermentation products for biological phosphorus removal. For wastewaters with a relatively high TKN:BOD ratio, the nitrate-nitrogen concentration in the return sludge may demand a high enough portion of the soluble BOD entering the anaerobic zone fermentation to result in less phosphorus removal.

In contrast, the anoxic stage of the UCT process is designed and operated to produce a very low nitrate-nitrogen concentration in recycle streams to the anaerobic fermentation zone. The recycle of mixed liquor from the anoxic stage to the anaerobic stage thereby provides optimum conditions for conversion of available soluble BOD to fermentation products. The mixed liquor recycle from the aerobic stage to the anoxic stage (recycle 2) can be controlled to assure a minimal nitrate-nitrogen concentration in recycle 1, while achieving some level of nitrogen removal in the anoxic zone.

A modified UCT process is also shown in Figure 6-6. In this case, the first anoxic zone is designed to reduce only the nitrate-nitrogen in the return activated sludge. The second anoxic zone is designed for a much higher quantity of nitrate-nitrogen removal as it receives mixed liquor recycled from the nitrification zone.

Another modification to the UCT process is the Virginia Initiative Process (VIP), developed and patented by CH2M HILL consultant engineers, but is available as a license-free process. This process provides multiple stages in the anoxic zone of the three stage anaerobic-anoxic-aerobic system and is operated at much higher loadings and lower solids retention times (SRTs) than the UCT process. The mixed liquor feed stream to the anaerobic zone is taken from the last anoxic zone stage.

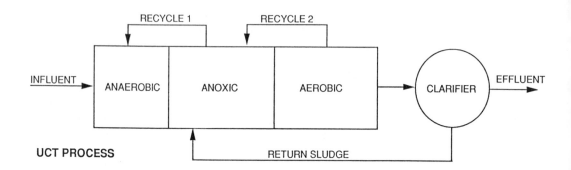

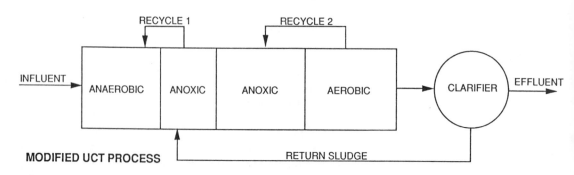

Figure 6-6. UCT process flow schematics.

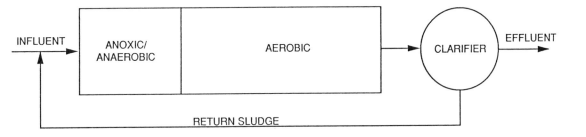

Figure 6-7. Operationally modified activated sludge system for biological phosphorus removal.

If conditions are favorable, operational changes can be made in existing activated sludge systems to create an anaerobic fermentation zone ahead of the aeration zone to promote biological phosphorus removal. Figure 6-7 indicates this approach. In practice, it typically involves turning off air flow or aerators in the front of the activated sludge basin. As described in Section 6.2, this technique was demonstrated during the earlier investigations of phosphorus removal with the Bardenpho process. Similarly, the plug flow plants in the United States, for which high levels of phosphorus removal were reported, had insufficient aeration at the front end of the aeration basins that inadvertently promoted the anaerobic-aerobic contacting sequence(5,6,7).

6.5 Factors Affecting Biological Phosphorus Removal Performance

There are many factors that can affect the phosphorus removal efficiency of these systems. These factors relate to wastewater characteristics, system design and operational methods. These factors can be divided into the following categories:

1. Environmental factors, such as DO, temperature, and pH.

2. Design parameters, such as system Solids Retention Time (SRT), anaerobic zone detention time, aerobic zone detention time, and waste sludge handling methods.

3. Substrate availability as affected by influent wastewater characteristics, the level of VFA production and the presence of nitrates.

The overall performance may also be affected by the effluent total suspended solids (TSS) concentration. Assuming a four percent phosphorus content in the mixed liquor effluent, TSS concentrations ranging from 10-20 mg/L would contribute an effluent particulate phosphorus concentration of 0.4 to 0.8 mg P/L. If the discharge standard is less than 1 mg/L as total phosphorus, a very low effluent soluble phosphorus concentration would be required. Effluent filtration to remove the phosphorus-containing TSS or chemical addition to lower the soluble phosphorus concentration may be necessary.

6.5.1 Environmental Factors

No specific studies have been undertaken to observe biological phosphorus uptake in the aerobic zone as a function of DO concentration. The early observations on phosphorus removal in the conventional plug flow systems suggest that with a DO concentration above 2 mg/L sufficient phosphorus uptake occurs provided that the aerobic detention time is long enough.

Biological phosphorus removal was studied in laboratory batch units over a temperature range of 5-15°C by Sell et al.(42). The amount of phosphorus removed at 5°C was 40 percent greater than that removed at 15°C. They attributed the improvement to a population shift to more slow growing psychrophilic bacteria that had a higher yield. Pilot studies at 5°C produced better phosphorus removal with an effluent soluble Pi of 0.9 mg P/L compared to when the system was operated at 15°C or above(43). The phosphorus content of the sludge of the 5°C operation was 4.7 percent compared to a range of 3.5 to 4.9 percent for higher temperature operating periods. A full-scale anaerobic-aerobic system evaluation at Pontiac, Michigan revealed that phosphorus removal was not affected by temperatures as low as 10°C (44).

Though phosphorus removal capacity does not seem to be affected by low temperature operations, Shapiro et al.(45) showed that the specific phosphorus release rate for a batch activated sludge sample increased by a factor of 5 as the temperature increased from 10°C to 30°C. This implies that more time may be required in the anaerobic zone at low temperatures for either fermentation to be complete and/or substrate uptake to occur.

Results of studies on the effects of pH suggest that more efficient biological phosphorus removal occurs at pH values from 7.5 to 8.0. Pure culture studies by Groenestijn and Deineman(46) showed that the maximum specific growth rate of **Acinetobacter** was 42 percent higher at a pH of 8.5 compared to a pH of 7.0. Tracy and Flammino(47) studied the effect of pH on the specific phosphorus uptake rate in the aerobic zone. There was little affect of pH between a range of 6.5 to 7.0. Below a pH of 6.5 activity steadily declined and all activity was lost at a pH of 5.2.

6.5.2 Design Parameters

Important design parameters for biological phosphorus removal systems are the system SRT, the anaerobic contact time and the aerobic detention time. The SRT value selected for design will be a function of treatment requirements and will increase as the system is designed for BOD removal, nitrification, or nitrification-denitrification. Longer SRT designs result in lower sludge production, which results in a lower amount of biological phosphorus removal, since the phosphorus is removed with the waste sludge.

Figure 6-8 illustrates the effect of SRT on phosphorus removal capacity, assuming two different mixed liquor phosphorus contents. For longer SRT designs with lower percent phosphorus contents, more BOD must be removed per unit of phosphorus removed. For example, assuming a 4.5 percent waste activated sludge phosphorus content, about 33 mg of BOD is required for each mg of phosphorus removed at an SRT of 25 days; a BOD:P ratio of 33:1. At the same sludge phosphorus content, the BOD:P ratio decreases to 25:1, using an eight day SRT. Fukase et al.(23) found in an anaerobic-aerobic pilot plant system treating municipal wastewaters that the BOD:P removal ratio increased from 19 to 26 as the SRT was increased from 4.3 to 8.0 days. At the same time the phosphorus content of the activated sludge decreased from 5.4 to 3.7 percent.

These results indicate that system designs requiring longer SRTs need a greater amount of BOD removal to meet low effluent phosphorus concentrations. Thus, a Bardenpho system has less phosphorus removal capacity than a Phoredox system for the same influent BOD concentration. This analysis assumes that the phosphorus removing organisms are not affected by SRT, which at present has not been verified or refuted.

The anaerobic contact time for biological phosphorus removal systems has in most cases been arbitrarily selected between 1-2 hours. The detention time is needed to allow sufficient fermentation to provide VFA for uptake by the phosphorus-storing organisms. The VFA uptake rate may also be important when considering the size of the anaerobic contactor, but measurements for VFA in the anaerobic zone suggests that it is taken up as fast at it is produced.

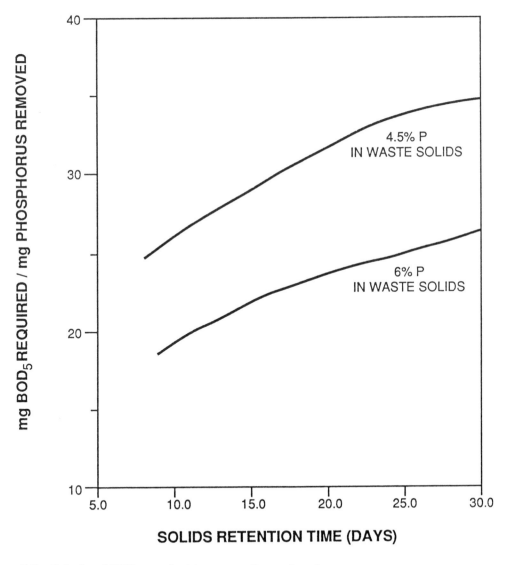

Figure 6-8. Calculated BOD_5 required to remove 1 mg phosphorus.

Figure 6-9, developed from cultures batch fed with acetate, shows that the organic uptake rate is a function of organic loading to the anaerobic zone. It further shows that at about two hours most of the applied COD is removed from solution. These data also suggest that an anaerobic zone divided into discrete stages would result in a more rapid uptake of organics and could be designed with a smaller volume than a single completely mixed anaerobic zone. However, the cost savings due to the smaller reactor volume would have to be weighed against added costs for additional mixers and divider walls.

Barnard(48) cautions against having too long of an anaerobic zone contact time. He points out that phosphorus release may occur under such conditions without the uptake of VFA compounds. When this occurs there are not sufficient carbon storage products within the cell to produce enough energy to force full uptake of the released phosphorus during the aerobic contact period. He terms this phosphorus release a "secondary release."

The aerobic tank is important for maintaining conditions for soluble phosphorus uptake after its release in the anaerobic zone. As these reactors are designed to provide a sufficient aerobic detention time for nitrification or BOD removal, sufficient time is expected for biological phosphorus uptake. This is a more critical issue if the aerobic tank is not fully oxygenated at all times. No definitive field study has been done at this time to evaluate the aerobic detention time, though some batch studies indicate that 1-2 hours is sufficient. Comeau(22) shows that the rate of phosphorus uptake in the aerobic zone increases as the level of organic storage products is increased. He observed phosphorus uptake rates generally ranging from 10-30 mg P/hr-L. Since soluble Pi release levels are generally in the range of 20-40 mg P/L in the anaerobic zone, a fully aerobic detention time between 1-2 hours appears to be a feasible operating range.

Since significant phosphorus can be released when the mixed liquor of a biological phosphorus removal system is subjected to anaerobic conditions, care must be taken during sludge processing. Dissolved air flotation must be used instead of gravity thickening. Recycle streams from solids dewatering processes and digestion processes must be carefully evaluated. In one example, use of sludge drying beds at the Palmetto, Florida Bardenpho facility resulted in insignificant levels of released phosphorus in the underdrain return(16). The nutrient rich sludge from this application is also used as a fertilizer.

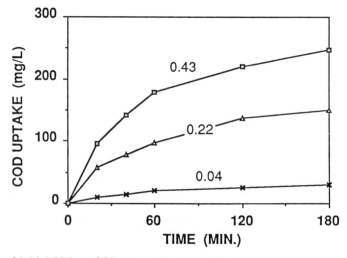

Figure 6-9. Effect of initial F/M on COD uptake in anaerobic zone.

6.5.3 Substrate Availability

The basic biological phosphorus removal mechanism model shows the importance of having organic fermentation products available for the phosphorus-storing organisms. The greater the amount of acetate and propionate made available in the anaerobic zone, the greater will be the amount of phosphorus removal. Since the fermentation products in the anaerobic zone are assimilated by the phosphorus-storing microorganisms about as fast as they are produced, it is not possible to directly measure the fermentation substrate available in a given wastewater. This makes it difficult to define the phosphorus removal capacity for a given wastewater. Attempts have been made to relate the effluent soluble phosphorus from a system to the amount of BOD added relative to the influent phosphorus concentration. An example of this is shown in Figure 6-10. A major problem with this evaluation is that only total BOD data were available and not soluble BOD data. The BOD component susceptible to fermentation in the short detention time anaerobic zone is the soluble BOD. Another short coming of the data is that the data also are affected by the varied SRTs and nitrate concentrations present in the systems. The data suggest that a total BOD/P ratio in the range of 20-30 may provide effluent soluble phosphorus concentrations below 1 mg P/L for systems with relatively low SRTs.

Hong et al.(18) have recommended an influent soluble BOD/P ratio of at least 15:1 to achieve low effluent soluble phosphorus concentrations in relatively short SRT anaerobic-aerobic systems. Siebritz et al.(49) attempt to identify the amount of fermentation products that can be produced by defining a "readily degradable" component of the influent municipal wastewater. They proposed measuring this portion of the influent BOD by oxygen consumption measurements after contacting a sample with mixed liquor. Nicholls et al.(50) prefer to use nitrate consumption as an indicator of this substrate, since it is easier to quantify the amount used.

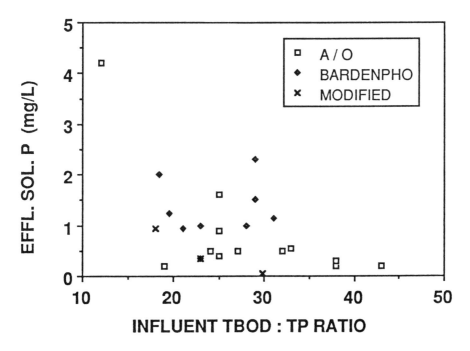

Figure 6-10. Effluent soluble phosphorus concentration vs. influent TBOD:TP ratio.

The introduction of nitrate into the anaerobic zone can deplete the readily available substrate supply for the biological phosphorus-removing organisms. Assuming a microbial yield of 0.3 g VSS/g COD removed, the following shows the amount of BOD demanded for each gram of nitrate-nitrogen that would then not be available for VFA production.

1. A reasonable biomass yield is:

$$Y = 0.3 \text{ g VSS/g COD}_{removed},$$

2. Assuming that the oxygen equivalent of the biological VSS is 1.42 g O_2/ g VSS, the fraction of $COD_{removed}$ that goes to cell production can be estimated:

Fraction of COD to cell mass $= (1.42 \text{ g } O_2/\text{g VSS}) (0.3 \text{ g VSS/g COD}_{removed})$

$= 0.43 \text{ g } O_2 \text{ as cells/g COD}_{removed}$

3. The fraction of COD that is not accounted for by cell mass represents oxidation:

The oxygen used for oxidation $= 1.0 \text{ g } O_2/\text{g COD}_{removed} - 0.43 \text{ g } O_2 \text{ as cells/g COD}_{removed}$

$= 0.57 \text{ g } O_2/\text{g COD}_{removed}$

The nitrate-nitrogen used to supply an equivalent amount of oxygen:

$= (0.57 \text{ g } O_2/\text{g COD}_{removed}) / (2.86 \text{ g } O_2 \text{ equiv/g NO}_3\text{-N})$

$= 0.20 \text{ g NO}_3\text{-N/g COD}_{removed}, \text{ or } 5 \text{ g COD}_{removed}/\text{g NO}_3\text{-N}$

This means 5 g COD (equivalent to about 3.4 g BOD_5) may be used for each g of NO_3-N added to the anaerobic zone.

Thus for a relatively weak wastewater, nitrate entering the anaerobic zone can significantly deplete the BOD available for conversion to anaerobic fermentation products. This will then decrease phosphorus removal efficiency or even prevent biological phosphorus removal depending on the amount of nitrate received.

For wastewaters with high soluble organic concentrations, the effect of nitrate may not be significant. At high enough soluble organic levels nitrate reduction, VFA reduction, and phosphorus release can occur simultaneously(25). The fermentation and phosphorus release is likely occurring inside the biological floc where nitrate is not available due to its depletion at the external layers of the floc.

6.5.4 Phosphorus Removal Versus VFA Production

The amount of phosphorus that can be removed per unit of VFA or acetate generated in or added to the anaerobic zone is a function of the cell yield and net amount of phosphorus stored in the wasted biological mass. If this were known as well as the amount of BOD that may be converted to acetate (HAC), the phosphorus removal capacity for a given wastewater could be predicted.

A typical cell yield for **Acinetobacter** is 0.40 g VSS/g HAC. Assuming a cell phosphorus content of 10 percent, 0.04 g phosphorus can be removed per g of HAC used or the removal of 1 g of phosphorus requires 25 g of HAC. This would yield an approximate BOD:P ratio of 17:1 which compares well with Hong's(18) estimate of 15:1. In contrast to this, work by Wentzel et al.(51) showed that 1 g of phosphorus could be removed with each addition of 8.9 g HAC. This result suggests a cellular phosphorus content of 28 percent for the phosphorus-storing microorganisms instead of the 10 percent assumed above.

Parallel biological phosphorus removal pilot plants were operated by Comeau(22). One of the units was fed an additional amount of VFAs consisting of 63 percent acetate and 37 percent propionate. The unit receiving the VFA addition showed an increase of 1 g phosphorus removal for every 6.4 g VFA added. Similar effects were observed at the full-scale Bardenpho facility at Kelowna, British Columbia, Canada(52). With the addition of VFA to one train, the effluent soluble phosphorus decreased from 2 mg P/L to 0.5 mg P/L. One g of phosphorus removal was observed for each 6.7 g VFA added. These results suggest that biological phosphorus storing organisms may show enhanced phosphorus uptake with VFA addition in the range of 1 mg phosphorus for each 7 to 9 mg of HAC added.

6.6 Improving Biological Phosphorus Removal

Experience with the operation of full-scale biological phosphorus removal systems shows that effluent total phosphorus concentrations of less than 1 mg P/L are not always achieved. In many cases some chemical addition is necessary to meet effluent phosphorus limits. Based on the discussion in the preceding section, there may not be enough soluble, readily degradable organics in a wastewater influent that can be fermented to VFAs in the anaerobic zone to promote high enough levels of phosphorus removal. For many municipal wastewaters, the soluble BOD may only account for 40-60 percent of the total BOD in the wastewater. Thus a considerable organic resource exists in the particulate organic fraction of the wastewater that may be convertible to VFAs.

Attempts have been made to improve the production of VFAs by employing operating conditions that improved the fermentation of available BOD. Osborn and Nicholls(15) suspected the importance of VFAs for biological phosphorus removal and operated a primary digester with high loadings to encourage only acid fermentation. The fermented sludge was then fed to the anaerobic zone of a modified Bardenpho system which led to phosphorus removal improving. Eventually methane fermentation developed in the digester causing a decrease in the VFA production and phosphorus removal efficiency. Primary clarifier sludge was removed and fermented in separate tanks at the Kelowna Bardenpho facility. The VFA concentration of the fermenter effluent was 110-140 mg/L and resulted in a VFA concentration in the system influent of 9-10 mg/L. As shown earlier this decreased the effluent phosphorus concentration to 0.5 mg P/L.

Figures 6-11 and 6-12 show two fermentation tank designs that can be used with primary treatment to increase the availability of VFAs for phosphorus storing organisms. One uses a deep tank primary clarifier and the other uses a separate fermenter, which can also be used to thicken the sludge. Recycle of sludge around the fermenter provides for more efficient solids conversion and for release of fermentation VFAs into the primary clarifier effluent.

The use of a primary sludge fermenter to improve biological phosphorus removal treatment efficiency is increasing most notable for facilities in South Africa and Canada. A plant modification has been made to provide primary sludge fermentation for a Bardenpho facility in Payson, Arizona.

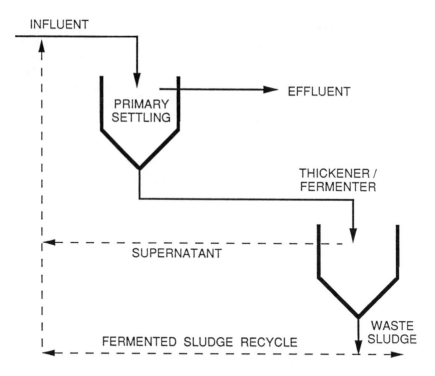

Figure 6-11. Primary sludge fermentation design

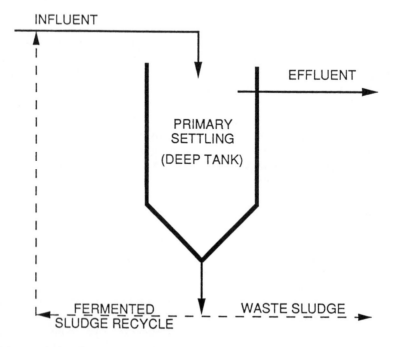

Figure 6-12. Primary sludge fermentation design.

Fundamental research on primary sludge fermentation by Eastman and Ferguson(53) showed that a solids retention time of about three days is necessary for maximum conversion of fermentable material to VFAs. They showed that only about 30 percent of the primary solids could be converted to VFAs and that a large fraction of the solids (about 40 percent) were lipids that could not be broken down under acid fermentation conditions. The conversion to VFA is also affected by the level of methanogenic activity that may develop in the fermentation reactor. This will be a function of pH, temperature, and how well the fermenter SRT is maintained at low levels. Less organic loss to methane was observed at pH values in the range of 5.0 to 6.0, but at this lower range a higher percentage of propionic acid was produced.

Primary sludge fermentation studies reported by Rabinowitz and Oldham(54) showed that an SRT range of 3.5 to 5.0 days was optimal. Less VFA production was observed for operations at a 10 day SRT due to methane production. A 9% conversion of primary solids to VFA on a COD basis was achieved under the latter conditions versus a 30% maximum conversion estimated by Eastman and Ferguson. The lower conversion levels could be due to less optimal mixing, methane production, or to the characteristics of the solids.

Primary clarifiers should be operated to maximize sludge thickening in order to minimize the volume of the tank required for primary solids fermentation. Assuming a 5 percent primary solids underflow and a 5 day fermenter SRT, the equivalent detention time of the fermenter, based on the influent flow rate, could be as low as 0.50 hours. Thus the tank size could be modest compared to the other tanks for a facility.

General performance results from biological phosphorus removal systems suggest that soluble phosphorus concentrations of less than 0.5 mg/L can be expected if either chemicals are added or primary solids fermentation is employed to generate enough acetate to cause the additional removal of about 2 mg/L of phosphorus. This is a conservative estimate for wastewaters of average or moderately weak organic strength. Based on the relationships presented between additional phosphorus removal and VFA production, a VFA production of 10-14 mg/L, based on influent flow, would be required. Assuming a 30% BOD removal efficiency in the primary clarifier, the particulate degradable COD removed and available for fermentation would be about 90 mg/L based on the influent flow rate. Assuming a 20% conversion to VFA in the fermenter, a VFA production of about 18 mg/L is feasible. Thus, this analysis indicates that there can be enough additional VFA potential in the particulate BOD fraction of the influent wastewater to enhance the treatment capacity of biological phosphorus removal systems.

Long-term results at the Kelowna, B.C. plant support the above analysis. With primary fermentation, effluent total phosphorus concentrations of about 0.5 mg P/L have been routinely achieved. Since September 1989 the plant has added alum at a dosage of about 8 mg/L before the final clarifier to further improve performance. For the period from September 1989 to August 1990, the effluent total phosphorus averaged less than 0.20 mg P/L.

Another example of using primary sludge fermentation is the Nutrification Sludge Process, shown in Figure 6-13. This process was developed at the Orange Water and Sewer Authority (OWASA) Mason Farm wastewater treatment facility (8 MGD) near Chapel Hill, North Carolina(55). Pilot plant testing efforts showed that biological phosphorus removal using the either the A/O, Bardenpho, or UCT process was not feasible because of the low BOD in the trickling filter effluent entering the biological nutrient removal stage. The plant modification shown provided an anaerobic contact zone in the sludge recycle stream and VFAs for the biological phosphorus removal organisms were provided by fermentation of the primary solids. Since a significant concentration of nitrate-nitrogen exists in the sludge recycle stream, the first contact basin is an anoxic basin prior to the anaerobic reaction period.

161

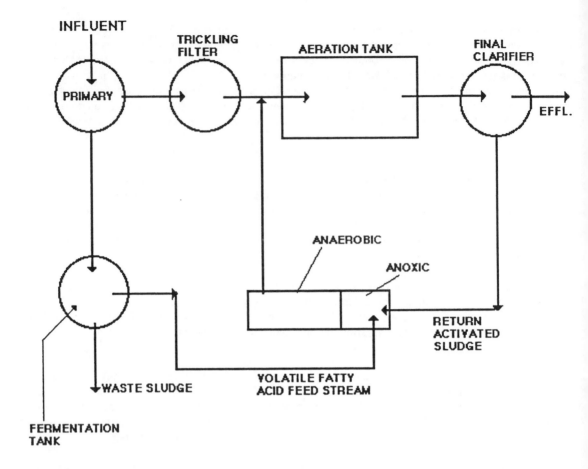

Figure 6-13. Nutrification Sludge Process.

Table 6-3 shows reported phosphorus concentrations for the Nutrification Sludge Process.

Table 6-3. Operations data for the Nutrification Sludge Process (monthly averages).

Parameter	January 1990	June 1990
Ave. daily flow, MGD	6.0	5.3
Wastewater temperature, °C	15	24
Influent total P, mg/L	7.1	6.3
Effluent total P, mg/L	0.7	0.6

162

Fundamental data and full-scale plant evaluations indicate that fermentation of primary solids can provide a VFA stream of sufficient strength to enhance biological phosphorus removal. Further improvement is also possible with a small dosage of alum before the clarifier. Effluent total phosphorus concentrations in the range of 0.2-0.5 mg P/L are possible with this technology compared to effluent total phosphorus concentrations in the range of 1-3 mg P/L for systems without primary solids fermentation or chemical addition.

6.7 References

1. Greenburg, A. E., G. Levin, and W. J. Kauffman. Effect of phosphorus removal on the activated sludge process. **Sewage and Industrial Wastes, 27**, 227, 1955.

2. Srinath, E. G. *et al*. Rapid removal of phosphorus from sewage by activated sludge. **Experientia** (Switzerland), **15**, 339, 1959.

3. Alarcon, G. O. Removal of phosphorus from sewage. Master thesis, Johns Hopkins University, Baltimore, MD, 1961.

4. Levin, G. V. and J. Shapiro. Metabolic uptake of phosphorus by wastewater organisms. **Jour. Water Pollut. Control Fed., 37**, 800, 1965.

5. Vacker, D. *et al*. Phosphate removal through municipal wastewater treatment at San Antonio, Texas. **Jour. Water Pollut. Control Fed., 39**, 750, 1967.

6. Bergman, R. D. *et al*. Continuous studies in the removal of phosphorus by the activated sludge process. **Chem. Engr. Prog. Symp. Ser., 67**, 117, 1970.

7. Milbury, W. F. *et al*. Operation of conventional activated sludge for maximum phosphorus removal. **Jour. Water Pollut. Control Fed., 43**, 1890, 1971.

8. Menar, A. B. and D. Jenkins. The fate of phosphorus in waste treatment processes: The enhanced removal of phosphate by activated sludge. Proceedings of the 24th Purdue Industrial Waste Conference, Lafayette, Indiana, 1969.

9. Barnard, J. L. Cut P and N without chemicals. **Water and Wastes Engineering, 7**, 1974.

10. Barnard, J. L. A review of biological phosphorus removal in the activated sludge process. **Water SA, 2**, 136, 1976.

11. Simpkins, M. J., and A. R. McLaren. Consistent biological phosphate and nitrate removal in an activated sludge plant. **Progr. Water Technol. (G.B.), 10**(5/6), 433, 1978.

12. Nichols, H. A., and D. W. Osborn. Bacterial stress: Prerequisite for biological removal of phosphorus. **Jour. Water Pollut. Control Fed., 51**(3), 557, 1979.

13. Nicholls, H. A. Full scale experimentation on the new Johannesburg extended aeration plants. **Water S.A., 1**, 121, 1975.

14. Venter, S. L. V. *et al*. Optimization of the Johannesburg Olifantsvlei extended aeration plant for phosphorus removal. **Prog. in Wat. Technology, 10**, 279, 1978.

15.   Osborn, D. W., and H. A. Nicholls. Optimization of the activated sludge process for the biological removal of phosphorus. Int. Conf. on Advanced Treatment and Reclamation of Wastewater, Johannesburg, South Africa, June 1977.

16.   Stensel, H. D. *et al.* Performance of first U.S. full scale Bardenpho facility. Proceedings of EPA International Seminar on Control of Nutrients in Municipal Wastewater Effluents, San Diego, CA, September 1980.

17.   Hong, S. N. *et al.* A biological wastewater treatment system for nutrient removal. Presented at the 54th Annual WPCF Conference, Detroit, Michigan, October 4-9, 1981.

18.   Hong, S. N. *et al.* A biological wastewater treatment system for nutrient removal. Presented at EPA Workshop on Biological Phosphorus Removal in Municipal Wastewater Treatment, Annapolis, Maryland, June 22-24, 1982.

19.   Timmerman, M. W. Biological phosphate removal from domestic wastewater using anaerobic/aerobic treatment. Chapter 26 in **Development in Industrial Microbiology,** 285, 1979.

20.   Deinema, M. H. *et al.* The accumulation of polyphosphate in acinetobacter spp. Microbial Letters, Federation of Microbiological Societies, 273-279, 1980.

21.   Buchan, L. The location and nature of accumulated phosphorus in seven sludges from activated sludge plants which exhibited enhanced phosphorus removal. **Water SA,** 7, 1, 1981.

22.   Comeau, Y. The role of carbon storage in biological phosphate removal from wastewater. Ph.D. Thesis, University of British Columbia, Vancouver, Canada, March 1989.

23.   Fukase, T., M. Shibeta, and X. Mijayi. Studies on the mechanism of biological phosphorus removal. **Japan Journal Water Pollution Research,** 5, 309, 1982.

24.   Arvin, E. Biological removal of phosphorus from wastewater. **Environmental Control,** CRC Critical Review, 15, 25-69, 1985.

25.   Rabinowitz, B. The role of specific substrates in excess biological phosphorus removal. Ph.D. Thesis, The University of British Columbia, Vancouver, British Columbia, Canada, October 1985.

26.   Wentzel, M. C., P. L. Dold, G. A. Ekama, and G. v. R. Marais. Kinetics of biological phosphorus release. Enhanced Biological Phosphorus Removal from Wastewater, Vol. 1, IAWPRC Post Conference Seminar, p. 89, September 24, 1984, Paris, France.

27.   Potgieter, D. J., and B. W. Evans. Biological changes associated with luxury phosphate uptake in a modified Phoredox activated sludge system. **Water Sci. Technol.,** 15(3/4), 105-115, 1983.

28.   Siebritz, I. P., G. A. Ekama and G. v. R. Marais. A parametric model for biological excess phosphorus removal. **Water Sci. Technol.,** 15(3/4), 127-152, 1983.

29. Comeau, Y., W. K. Oldham and K. J. Hall. Dynamics of carbon reserves in biological dephosphatation of wastewater. Adv. in Water Pollution Control, Proc. Rome Specialist Conf. on Biological Phosphate Removal from Wastewater, 39-56, 1987.

30. Gerber, A., E. S. Mostert, C. T. Winter and R. H. de Villiers. Interactions between phosphate, nitrate and organic substrate in biological nutrient removal processes. **Water Sci. Technol.**, **19**, 183-194, 1986.

31. Harold, F. M. Inorganic polyphosphates in biology: Structure metabolism and function. **Bacteriol. Reviews**, **30**, 772, 1966.

32. Comeau, Y., K. J. Hall, R. E. W. Hancock and W. K. Oldham. Biochemical model for biological enhanced phosphorus removal. **Water Res.**, **20**, 1511-1521, 1986.

33. Fuhs, G. W., and M. Chen. Microbial basis for phosphate removal in the activated sludge process for the treatment of wastewater. **Microb. Ecol.**, **2**, 119, 1975.

34. Lotter, L. H., and M. Murphy. The identification of heterotrophic bacteria in an activated sludge plant with particular reference to polyphosphate accumulation. **Water SA**, **11**(4), 172, October 1985.

35. Comeau, Y., B. Rabinowitz, K. J. Hall, and W. K. Oldham. Phosphate release and uptake in enhanced biological phosphorus removal from wastewater. **Jour. Water Pollut. Control Fed.**, **59**, 707, 1987.

36. Rensink, J. H., H. J. G. W. Donker, and H. P. de Vries. Biological P removal in domestic wastewater by the activated sludge process. Presented at 5th European Sewage and Refuse Symposium, Munich, West Germany, Procs. 487-502, June 1981.

37. Levin, G. V., G. J. Topol, and A. G. Tarnay. Operation of full scale biological phosphorus removal plant. **Jour. Water Pollut. Control Fed.**, **47**(3), 1940, 1975.

38. Arora, M. L., Barth, E. F. and M. B. Umphres. Technology evaluation of sequencing batch reactors. **Jour. Water Pollut. Control Fed.**, **57**, 807, 1985.

39. Irvine, R. L. *et al*. Municipal application of sequencing batch treatment at Culver, Indiana. **Jour. Water Pollut. Control Fed.** 55, 484, 1983.

40. Irvine, R. L. *et al*. Organic loading study of full-scale sequencing batch rectors. **Jour. Water Pollut. Control Fed.**, **57**, 847, 1985.

41. Ekama, G. A., Marais, G. v. R. and I. P. Siebritz. Biological excess phosphorus removal. Chapter 7 in **Theory, Design, and Operation of Nutrient Removal Activated Sludge Processes**, Water Research Commission, Pretoria, South Africa, 1984.

42. Sell, R. L. *et al*. Low temperature biological phosphorus removal. Presented at the 54th Annual WPCF Conference, Detroit, Michigan, October 1981.

43. Vinconneau, J. C., Hascoet, M. C. and M. Florentz. The first applications of biological phosphorus removal in France. Proceedings of the International Conference, Management Strategies for Phosphorus in the Environment, Lisbon, Portugal, July 1-4, 1985.

44. Kang, S. J. *et al*. A year's low temperature operation in Michigan of the A/O system for nutrient removal. Presented at the 58th Annual Water Pollution Control Federation Conference, Kansas City, Missouri, October 1985.

45. Shapiro, J., G. V. Levin, and Z. G. Humberto. Anoxically induced release of phosphate in wastewater treatment. **Jour. Water Pollut. Control Fed.**, **39**, 1810, 1967.

46. Groenestijn, J. W., and M. H. Deinema. Effects of cultural conditions on phosphate accumulation and release by acinetobacter strain 210A. Proceedings of the International Conference, Management Strategies for Phosphorus in the Environment, Lisbon, Portugal, July 1-4, 1985.

47. Tracy, K. D., and A. Flammino. Kinetics of biological phosphorus removal. Presented at the 58th Annual Water Pollution Control Federation Conference, Kansas City, Missouri, October 1985.

48. Barnard, J. L. Activated primary tanks for phosphate removal. **Water SA**, **10**(3), July 1984.

49. Siebritz, I. P., G. A. Ekama, and G. v. R. Marais. Biological phosphorus removal in the activated sludge process. Research Report W46, Department of Civil Engineering, University of Capetown, South Africa, 1983.

50. Nicholls, H. A., A. R. Pitman, and D. W. Osborn. The readily biodegradable fraction of sewage; its influence on phosphorus removal and measurement. Enhanced Biological Phosphorus Removal from Wastewater, Vol. 1, IAWPRC Post Conference Seminar, September 24, 1984, Paris, France, 105.

51. Wentzel, M. C., R. E. Loewenthal, G. A. Ekama and G. v. R. Marais. Enhanced polyphosphate organism cultures in activated sludge systems - Part 1: Enhanced culture development. **Water SA**, **14**, 81-92, 1988.

52. Oldham, W. K. and G. M. Stevens. Operating experiences with the Kelowna pollution control centre. Proceedings of the Seminar on Biological Phosphorus Removal in Municipal Wastewater Treatment, Penticton, British Columbia, Canada, April 17 and 18, 1985.

53. Eastman, J. A., and J. F. Ferguson. Solubilization of particulate organic carbon during the acid phase of anaerobic digestion. **Jour. Water Pollut. Control Fed.**, **53**, 352, 1981.

54. Rabinowitz, B., and W. K. Oldham. The use of primary sludge fermentation in the enhanced biological phosphorus removal process. **Proc. New Directions and Research in Waste Treatment and Residuals Management**, University of British Columbia, Vancouver, B.C., 347, 1985.

55. Kalb, K., R. Williamson, and M. Frazier. Nutrified Sludge: An innovative process for removing nutrients from wastewater. Presented at the Annual Water Pollution Control Federation Conference, Washington, D.C., 1990.

Chapter 7

Design and Operation of Biological Phosphorus

Removal Facilities

7.1 Process Options

The basic principles of biological phosphorus removal are described in Chapter 6 of this manual. All biological phosphorus removal systems utilize these same basic biochemical principles, which may be summarized in the following two-step process description:

1. Certain microorganisms, when subjected to anaerobic conditions, assimilate and store fermentation products produced by other facultative bacteria. These microorganisms derive energy for this assimilation from stored polyphosphates, which are hydrolyzed to release energy. The resulting phosphorus is released to the mixed liquor.

2. These same microorganisms, when subsequently exposed to aerobic conditions, consume both phosphorus (which is used for cell synthesis and stored as polyphosphates) and oxygen to metabolize the previously stored substrate for energy production and cell synthesis.

Once the phosphorus is stored in the microorganisms, it is important that the sludge is not inadvertently subjected to anaerobic conditions in subsequent treatment steps, such as a secondary clarifier. If anaerobic conditions develop, the phosphorus may again be released into solution. Phosphorus removal ultimately occurs in these systems through its removal from the system in the waste activated sludge(WAS).

The following paragraphs describe the various process options currently available for biological phosphorus removal. Many of the biological phosphorus removal processes are patented and require the payment of license fees for their use.

7.1.1 Phosphorus Removal Only

Two common biological phosphorus removal processes are marketed in the U.S.A. for removal of phosphorus only, and not nitrogen: (1) A/O, and (2) Phostrip. These processes are not generally capable of achieving significant nitrogen removal. They may not be useful in applications requiring nitrification due to the interference to phosphorus removal caused by the presence of nitrate-nitrogen in the return activated sludge(RAS).

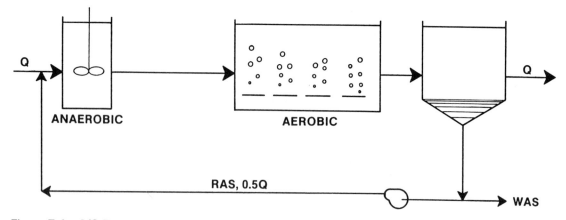

Figure 7-1. A/O Process.

A/O Process. The A/O process, which stands for anaerobic/oxic, is patented and marketed by Air Products and Chemicals, Inc.(1). A schematic of the process is shown in Figure 7-1. The A/O process is one of the simplest of the biological phosphorus removal systems, being very similar to a standard activated sludge process.

In this process, the mixed liquor (RAS plus influent flow to the secondary process) passes first through an anaerobic zone where the first of the process steps described above (phosphorus release) occurs. The mixed liquor then leaves the anaerobic zone and passes directly through an aerobic zone where the second reaction (phosphorus uptake) occurs. Following aeration, the mixed liquor passes to a secondary clarifier where the phosphorus-enriched sludge is settled from the process and returned to the anaerobic zone.

As with an activated sludge plant, a portion of the sludge is removed from the system, or wasted, as necessary to maintain the desired mean cell residence time (MCRT) or solids retention time (SRT). The wasted sludge contains the phosphorus removed by the process. Because the anaerobic zone is located in the main liquid process stream, A/O is referred to as a mainstream biological phosphorus removal process.

The most notable characteristic of the A/O process is its high rate operation. The process uses a relatively short SRT and high organic loading rates, resulting in increased sludge production rates and phosphorus removal rates relative to the Modified Bardenpho process described below. This results in a high removal of phosphorus per unit of BOD as compared to the other mainstream biological phosphorus removal processes described below.

Phostrip Process. The Phostrip process, originally developed by Levin in 1965(2), includes both biological and chemical methods for phosphorus removal. A schematic of the process is shown in Figure 7-2.

The main plant flow stream for the Phostrip process is essentially an activated sludge process, comprised of an aerobic zone, or aeration basin, and a secondary clarifier. The phosphorus removal treatment process receives a sidestream flow diverted from the RAS flow stream. This sidestream sludge flow, typically 10 to 30 percent of the plant influent flow rate, is diverted to a stripper tank.

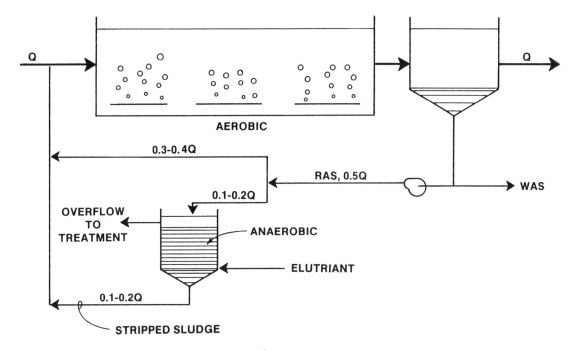

Figure 7-2. Phostrip Process.

Anaerobic conditions are maintained in the stripper tank, thereby encouraging the release of soluble phosphorus from the RAS microorganisms. Because anaerobic conditions are created not in the main liquid process stream but in the RAS sidestream, Phostrip is referred to as a sidestream process. The release of phosphorus in the stripper tank is thought to be caused by the following mechanisms: (1) release of phosphorus from the phosphorus-removing microorganisms during the uptake of fermentation products; and (2) release of phosphorus from lysed bacteria in the stripper(3). The average solids detention time in the stripper tank is generally 8 to 12 hours.

The soluble phosphorus is "washed" from the RAS solids by the continuous addition of elutriation water to the stripper. The stripped RAS flow stream is recombined with the remainder of the RAS flow stream to be returned to the activated sludge system. The elutriation water, on the other hand, typically flows from the stripper tank to a reactor clarifier. Lime is added to the flow stream to precipitate the phosphorus, as in standard chemical phosphorus treatment. The resulting sludge must then be disposed of, as described in Chapter 5. The supernatant from the clarifier is returned to the plant flow stream upstream of the secondary treatment process. As an alternative to the use of a separate reactor-clarifier, the phosphorus-rich supernatant from the stripper is sometimes dosed with lime and then recycled directly to the primary clarifier, where it settles with the primary sludge.

In addition to the phosphorus stripping and precipitation process described above, phosphorus is also removed from the wastewater in the Phostrip process through the WAS. The Phostrip process increases the phosphorus content of the WAS over that of a typical activated sludge system. This results in an increase of 50 to 100 percent in the phosphorus removed with the WAS, as compared to a standard activated sludge process(4).

169

The primary advantage of the Phostrip process is that it is not as sensitive to the $TBOD_5$ of the incoming wastewater as the other biological phosphorus removal processes. The Phostrip process is able to achieve less than a 1 mg/L effluent total phosphorus concentration in most cases. When compared to chemical phosphorus removal processes, the Phostrip process will generally require a lower chemical dosage than mainstream lime phosphorus removal, since a much smaller flow stream (10 percent of plant flow) is treated. As discussed in Chapter 4, the lime required for phosphorus precipitation is not stoichiometrically related to the quantity of phosphorus to be precipitated, but rather functions by raising the pH of the flow stream to allow precipitation of hydroxyapatite.

7.1.2 Combined Phosphorus/Nitrogen Removal

Many effluent permits require reduced levels of both phosphorus and nitrogen. In addition, biological processes for removal of these two contaminants may be incorporated into the standard activated sludge secondary treatment process with relative ease. These factors have encouraged the development of several combined processes for nitrogen and phosphorus removal. All of these processes include the same basic anaerobic/anoxic/aerobic components. However, significant differences exist among the processes with regard to the arrangement and number of these components, as well as the number and destination of recycle streams. All of these process options provide mainstream control of phosphorus.

The following paragraphs describe the following combined biological nutrient removal processes: (1) A^2/O, (2) Modified or Five-Stage Bardenpho, (3) UCT, and (4) VIP.

$\underline{A^2/O}$. The A^2/O process is a modification of the A/O process described previously. The acronym A^2/O stands for "anaerobic/anoxic/aerobic," which describes the basic process train shown schematically in Figure 7-3. As with the A/O process, the A^2/O process is proprietary, marketed by Air Products and Chemicals, Inc.

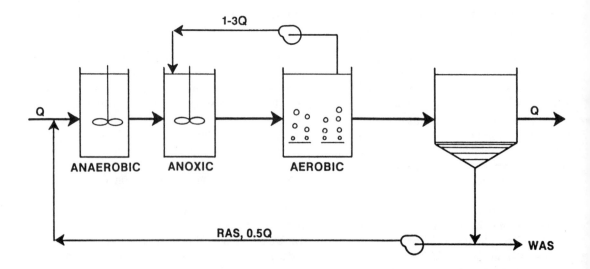

Figure 7-3. A^2/O Process.

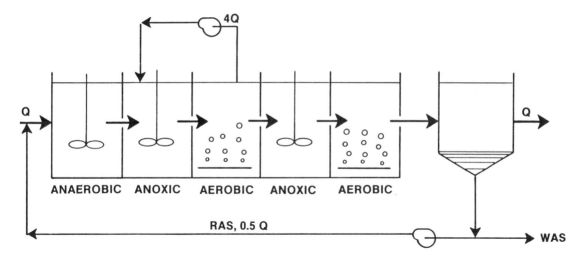

Figure 7-4. Five-Stage Bardenpho Process.

The A^2/O process is applicable to situations requiring only ammonia removal (nitrification) and those requiring nitrogen removal (nitrification/denitrification). Basically, the A^2/O process adds an anoxic zone between the anaerobic and aerobic zones already provided in the A/O process. This provides for nitrogen removal through denitrification, as described in Chapter 3. The anoxic zone is often included for ammonia removal only options to reduce the nitrate loading on the anaerobic zone through the RAS flow streams. Otherwise, the nitrate concentration in the RAS could reduce the efficiency and effectiveness of phosphorus removal.

Mixed liquor is recycled from the aerobic zone to the anoxic zone at a rate of 100 to 300 percent of the plant influent flow. Nitrogen removals of 40 to 70 percent have been achieved with the A^2/O process(3); phosphorus removal capability of the A^2/O process is somewhat less than that of the A/O process.

Modified or Five-Stage Bardenpho Process. In contrast to the high-rate operation of the A/O and A^2/O processes, the Five-Stage Bardenpho process is generally designed at low loading rates to improve the nitrogen removal performance of the system. The Bardenpho system is licensed and marketed in the U.S.A. by the Eimco Process Equipment Company. A schematic of the Five-Stage Bardenpho system is shown in Figure 7-4. This system is a modification of the original Bardenpho treatment scheme, in that an anaerobic zone is included at the beginning of the process train to provide for phosphorus removal in the system. The remainder of the process is essentially the same as the original Bardenpho system, with anoxic/aerobic/anoxic/aerobic zones in series and mixed liquor recycle from the first aerobic zone to the first anoxic zone. The RAS is recycled to the influent end of the anaerobic zone.

The nitrogen removal operational characteristics of the Bardenpho process are described in Chapter 3. The added anaerobic zone promotes the typical fermentation reactions and the substrate uptake/phosphorus release process described at the beginning of this chapter. The phosphorus uptake reaction then occurs in the first aerobic zone. The final aerobic zone serves the added purpose in the Five-Stage Bardenpho process of preventing the occurrence of anaerobic conditions in the secondary clarifiers and the associated release of phosphorus to the plant effluent. The SRT in the Five-Stage Bardenpho process is typically in the range of 10 to 20 days. Operating Five-Stage Bardenpho systems reportedly achieve total phosphorus concentrations in the effluent of 3 mg/L or less(3).

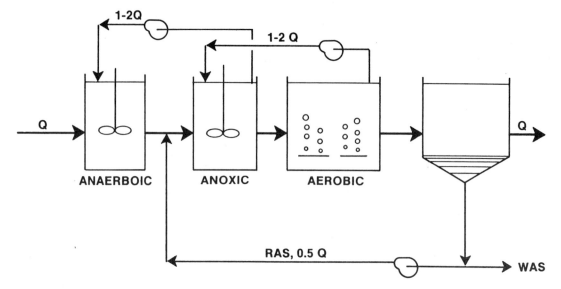

Figure 7-5. UCT Process.

University of Capetown (UCT). A further biological nitrogen and phosphorus removal process has been developed at the University of Capetown in South Africa(5). This process, named the University of Capetown (UCT) process, is shown schematically in Figure 7-5. The process includes the three basic zones, anaerobic/anoxic/aerobic, typical of the other biological phosphorus and nitrogen removal processes. The intent of this modified process is to reduce the nitrate loading on the anaerobic zone to optimize the phosphorus-related reactions in that zone. To accomplish this, the RAS is recycled to the anoxic zone instead of the anaerobic zone, and a second mixed liquor recycle flow is provided from the anoxic zone to the anaerobic zone. The anoxic zone is also operated to maintain a very low nitrate concentration in the zone and, hence, in the recycle stream. Otherwise the process functions similar to most of the other biological phosphorus removal processes, with phosphorus release in the anaerobic zone followed by excess phosphorus uptake in the subsequent aerobic zone.

In the UCT process, the recycle of nitrate from the aerobic reactor must be controlled so that the anoxic reactor is under loaded with nitrate to minimize the recycle of nitrate back to the anaerobic zone. Consequently, the nitrogen removal capacity of the process is not fully used. In order to resolve these two potential problems, a modification to the UCT process has been developed, as shown in Figure 7-6.

In this modification, the anoxic zone is separated into two zones. The first zone receives the RAS and provides recycle to the anaerobic zone. This first anoxic zone is therefore required to reduce only the nitrate in the RAS. The second anoxic zone receives the mixed liquor recycle flow, and it is the zone where the bulk of the denitrification occurs. By separating this zone from the zone feeding recycle to the anaerobic zone, excess nitrate may be recycled to the zone without jeopardizing the process. This essentially eliminates the problems identified above for the UCT process.

If the Modified UCT process is implemented, the recycle pumping system for the anaerobic zone should be designed to allow recycle from either of the anoxic zones. This would allow either the UCT or Modified UCT process to be used.

172

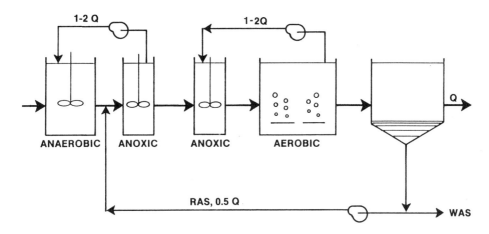

Figure 7-6. Modified UCT Process.

VIP Process. The Virginia Initiative Plant (VIP) process is another biological nitrogen and phosphorus removal process, which has been recently pilot tested and is now being implemented full-scale. This process is similar to the UCT process, and is shown schematically in Figure 7-7. It was developed for the expansion and upgrading of the Hampton Roads Sanitation District's Lamberts Point WWTP, as described later in this chapter.

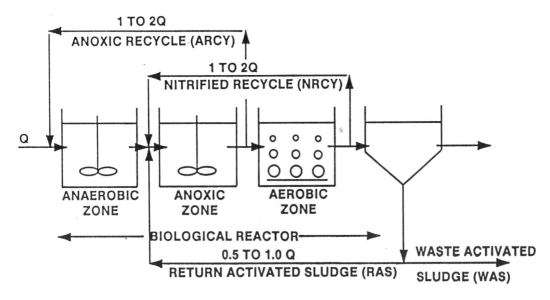

NOTE: A STAGED REACTOR CONFIGURATION IS
PROVIDED BY USING AT LEAST TWO COMPLETE
MIX CELLS IN SERIES FOR EACH ZONE OF THE
BIOLOGICAL REACTOR.

Figure 7-7. VIP Process.

Although the VIP process and the UCT process are similar, two significant differences exist:

1. Multiple complete mix cells are used in lieu of a single reactor for the anaerobic, anoxic, and aerobic treatment zones. The intent of this approach is to increase the rate of phosphorus uptake by virtue of a higher concentration of residual organics in the first aerobic cell.

2. A higher rate of operation is used to increase the proportion of active biomass in the mixed liquor. This more active biomass, with its greater phosphorus removal rate, reduces the necessary size of the reactors. The VIP process is designed for a total SRT of 5 to 10 days(6), while the UCT process is generally designed for an SRT of 13 to 25 days(5).

7.1.3 Sludge Fermentation

The concept of fermenting primary sludge to produce volatile fatty acids (VFAs) that can be used to enhance biological phosphorus removal was discussed in Chapter 6. VFAs appear to be the substrate used directly by the phosphorus removal bacteria. Conversion of primary sludge solids to VFAs and feeding of the generated VFAs to the anaerobic zone of a biological phosphorus removal process increases the mass of substrate available to the phosphorus removing bacteria. This, in turn, has proven effective in improving the performance of mainstream biological phosphorus removal facilities. Case histories documenting the benefits of VFA generation through fermentation of primary sludge, with subsequent feeding to the anaerobic zone, on the performance of biological phosphorus removal facilities are presented in Chapter 8.

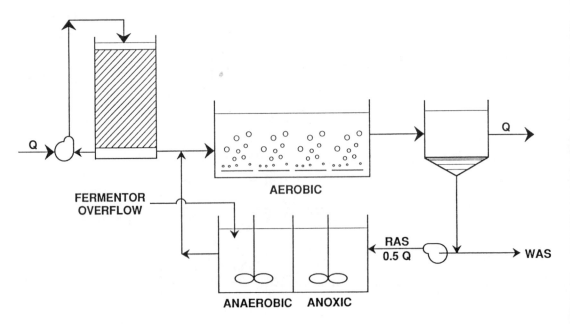

Figure 7-8. OWASA process.

VFAs can be fed to the anaerobic zone of any mainstream biological phosphorus removal system. Recently, process options that are dependent upon primary sludge fermentation have been developed. An example is the Orange Water and Sewerage Authority (OWASA) process, developed by OWASA for application at its full-scale plant in North Carolina. In this process, which is illustrated in Figure 7-8, effluent from an upstream trickling filter flows to an aeration basin (aerobic zone) where phosphorus uptake occurs. Mixed liquor from the aeration basin flows to the secondary clarifier where it is separated into a clear supernatant and a concentrated return activated sludge (RAS). RAS flows to a staged reactor which is mixed, but not aerated. Denitrification of any NO_x contained in the RAS occurs in the initial, anoxic stages of this reactor. Fermenter overflow, which is high in VFAs, is added to downstream, anaerobic stages of the reactor. Phosphorus release (and organism selection) occurs in the anaerobic portion of the RAS reactor. RAS then exits the RAS reactor and is mixed with the trickling filter effluent in the aeration basin where phosphorus uptake and nitrification occur. The incorporation of primary sludge fermentation into this process is necessary since the upstream trickling filter consumes most of the readily biodegradable organics in the influent wastewater.

7.2 Selection Factors

Several factors must be considered in the selection of the appropriate biological phosphorus removal process. These are described below.

7.2.1 Wastewater Characteristics

Two of the factors that affect the amenability of wastewater to biological phosphorus removal are: (1) $TBOD_5$/TP ratio, and (2) the content of readily biodegradable organic matter.

The $TBOD_5$/TP ratio is the ratio of $TBOD_5$ to total phosphorus (TP) of the wastewater entering the biological process. Research indicates that, for plants with a $TBOD_5$/TP ratio less than 20, it may be difficult to achieve an effluent total phosphorus level of 1.0 to 2.0 mg P/L if a mainstream treatment system (e.g., Bardenpho, A/O, UCT, etc.) is used(3). In contrast the Phostrip process is theoretically less sensitive to the influent wastewater strength. Therefore, it may be better suited for removing phosphorus from a weak wastewater. If nitrogen removal is also required for a wastewater with relatively typical nitrogen content, but relatively low in BOD (which precludes the use of the Phostrip process) the effluent phosphorus level from a mainstream system can be further reduced by chemical treatment and/or reducing the effluent TSS. Effluent TSS reduction can be accomplished either by conservative clarifier design or by effluent filtration.

As discussed in Chapter 6, the readily biodegradable organic content of a wastewater, particularly the VFA content, significantly affects the efficiency of a biological phosphorus removal system treating that wastewater. Phosphorus removal efficiency will be enhanced for wastewater with a relatively high content of readily biodegradable organic matter. Fermentation can be used to increase the VFA level of wastewaters containing low concentrations of readily biodegradable organic matter, as described above. Techniques for measuring the concentration of readily biodegradable organic matter of a particular wastewater have been developed(5). Preliminary assessments of the biodegradability of a subject wastewater can sometimes be made by experienced professionals. Septic wastewaters, in particular, may have a higher content of readily biodegradable organic matter than relatively fresh wastewater due to the fermentation which occurs in the collection system.

7.2.2 Nitrogen Removal Considerations

For many wastewaters, the selection of a phosphorus removal process is generally driven by the nitrogen removal requirements.

Nitrogen removal not required. If neither ammonia nor total nitrogen removal is required, it is unnecessary to use a phosphorus removal process that also removes nitrogen. These situations suggest use of the A/O or Phostrip processes, both of which are designed to remove only phosphorus.

The Phostrip process reportedly is capable of achieving an effluent total phosphorus level of 1 mg/L without additional treatment such as chemical addition or effluent filtration(3). The A/O process, on the other hand, may not achieve that level of treatment without process augmentation. The Phostrip process increases sludge production by virtue of the stripper effluent lime precipitation step. Also, the O&M costs for the Phostrip process may be greater than for the A/O process due to chemical requirements.

Nitrification or partial total nitrogen reduction. Some plants require ammonia removal (nitrification) only, or partial reduction of total nitrogen (TN) to a range of 6 to 12 mg/L. The processes which use a single anoxic zone, such as A^2/O, UCT, or VIP, are suitable for effluent nitrogen concentrations in this range.

As discussed earlier in this chapter, the A^2/O process recycles the RAS to the anaerobic zone. With nitrification occurring in the aerobic zone, it is likely that the RAS will contain a significant concentration of nitrate. Since the reduction of nitrate in the anaerobic zone utilizes substrate that would otherwise be stored by the phosphorus-removing organisms, the recycled nitrate may be thought of as effectively reducing the influent $TBOD_5$/TP ratio. If this ratio is already low (less than or equal to 20:1), the phosphorus removal efficiency of the process will be reduced(3). On the other hand, the UCT processes (UCT and Modified UCT) and the VIP process recycle the RAS to the anoxic zone. A denitrified anoxic mixed liquor stream is subsequently recirculated to the anaerobic zone. If operated properly, the nitrate concentration in the anoxic recycle can be maintained at or near zero. As a result, these processes do not adversely impact the influent $TBOD_5$/TP ratio.

In light of the potential impact of nitrate recycle to the anaerobic zone, the most important factor in selecting between the A^2/O process and the UCT and VIP processes is the influent $TBOD_5$/TP ratio. If the ratio is well above 20:1, the recycle of nitrate to the anaerobic zone may not be a problem. The A^2/O process may be more attractive for this case, since it does not include the additional anoxic mixed liquor recycle pumping requirement. If the $TBOD_5$/TP ratio of the influent to the biological system is in the range of 20:1 or less, the UCT or VIP processes should be considered over the A^2/O process to avoid further reduction in the $TBOD_5$/TP ratio and its negative impact on phosphorus removal efficiency.

Extensive Total Nitrogen Reduction Required. Many plants are faced with stringent effluent limits on both phosphorus and nitrogen. If low effluent nitrogen limits are imposed on a plant, along with moderate to low effluent phosphorus limits, the Five-Stage Bardenpho process is typically used. The Bardenpho system, with its two-stage anoxic/aerobic zones, is capable of consistently producing effluent nitrogen levels of 3 mg/L or less. It is also capable of achieving phosphorus removal. However, as with the A^2/O process, the phosphorus removal capabilities of the Modified Bardenpho process are adversely affected by a $TBOD_5$/TP ratio less than 20:1.

Table 7-1. Biological phosphorus removal process selection.

Process	Nitrification	Nitrogen Removal	Sensitivity to TBOD$_5$/TP Ratio
A/O	No	No[a]	Moderate
Phostrip	No	No[a]	Low
A^2/O	Yes	6-12 mg/L[b]	High
UCT	Yes	6-12 mg/L[b]	Low
VIP	Yes	6-12 mg/L[b]	Low
Bardenpho	Yes	3 mg/L[b]	High
Primary Sludge Fermentation	Yes or No	Varies	Low[c]

[a]Same degree of removal as achieved in conventional activated sludge facility.
[b]Approximate effluent concentration.
[c]Used, in particular, if wastewater is fresh and low in readily biodegradable organic matter.

7.2.3 Summary

Table 7-1 summarizes the effects of the factors identified above on biological phosphorus removal process selection. The primary factor affecting process selection is the degree of nitrification and/or nitrogen removal also desired for the process. If neither nitrification nor nitrogen removal are desired, then either the A/O or Phostrip process would typically be selected. Selection between these two processes requires more detailed analysis considering their relative costs, operability, discharge capability, and implementability. These considerations are relatively site-specific and must be addressed for each application.

If nitrification only or only a moderate degree of nitrogen removal (effluent total nitrogen of 6 to 12 mg/L) is desired, then either the A^2/O, UCT, or VIP process would typically be selected. Selection among these three options depends on projected wastewater characteristics (TBOD$_5$/TP ratio) and the degree of phosphorus removal capability required. The phosphorus removal capability of the A^2/O process is generally lower than that of the UCT or VIP process. The Bardenpho process would be selected when extensive nitrogen removal (effluent values of approximately 3 mg/L) is desired. Primary sludge fermentation can be utilized for wastewater with a relatively low content of readily biodegradable organic matter. The OWASA process, as illustrated in Figure 7-8, is an excellent example of the use of this procedure. While exceptions exist to the generalizations presented in Table 7-1, they represent a good starting point for preliminary biological process selection.

7.3 System Design

7.3.1 Process Design

The basic design considerations for each of the available biological phosphorus removal systems are discussed, as well as some general design considerations applicable to all systems. The discussion presented here is intended to be an overview. The reader is referred to the EPA **Design Manual for Phosphorus Removal**(3) for design procedures for the Phostrip, A/O, A^2/O, and Bardenpho processes; to the compilation of theory and design guidance by the University of Capetown(5) for the UCT-type processes; and to Daigger et al.(6) for the VIP process.

7.3.1.1 Sidestream Processes (Phostrip)

The mainstream portion of a Phostrip plant is a standard activated sludge process and is designed similar to other activated sludge facilities. As a result, the Phostrip process does not impact the selection of an SRT or food-to-microorganism (F/M) ratio for the activated sludge system design. The design concerns for the Phostrip process are: (1) size and configuration of the stripper tank; (2) size of the reactor-clarifier(s); (3) lime feed requirements; and (4) elutriation water source.

Stripper Tank Design. The stripper tank volume is based on the assumed RAS flow diverted to the unit, the necessary solids residence time (SRT) in the unit, and the influent and underflow sludge concentrations for the unit. The RAS feed rate is typically 20 to 30 percent of the plant influent flow. The SRT is generally in the range of 5 to 20 hours. The underflow is typically in the range of 10 to 20 percent of plant influent flow, indicating a thickening of 30 to 50 percent in the stripper(3).

The configuration of the unit, primarily the surface area, is determined based on the solids loading and an assumed solids flux rate. The stripper depth is based on the minimum required volume and the previously determined surface area, plus approximately 50 percent additional depth to allow solids inventory flexibility. The SRT achievable in the stripper is also considered. The stripper is typically 18 to 20 feet deep(3).

Reactor-Clarifier Design. The reactor-clarifier design is based on the flow of supernatant from the stripper and an assumed allowable overflow rate, in the same manner that a typical primary or secondary clarifier is designed. The supernatant flow from the stripper is comprised of two components: (1) water released from sludge thickening, and (2) elutriation water. An elutriation flow of 50 to 100 percent of the stripper feed flow is typically assumed(3). The reactor-clarifier design is typically based on an overflow rate of approximately 1,200 gallons per day per square foot (gal/d-ft^2) (3). A reactor-clarifier is used to provide an influent mixing zone for dispersing the lime into the stripper supernatant in a controlled manner.

Lime Feed Requirements. The lime feed rate is dependent on the characteristics of the supernatant that affect the ability of the lime to raise the pH of the stream to approximately 9 to 9.5. For most wastewaters, this requires a dose of 100 to 300 mg/L based on the supernatant flow rate.

Elutriation Water Source. Either primary effluent, secondary effluent, or supernatant from the lime precipitation reactor-clarifier are typically used as elutriation water sources. The quality of the elutriation water is an important factor in the efficient operation of the stripper. In general, the elutriation water should contain little, if any, dissolved oxygen, and it should not contain nitrate. The reduction of these substances will result in the consumption of a portion of the available substrate in the stripper. This substrate would then be unavailable for fermentation and assimilation in the phosphorus release process. A high-BOD elutriation water source is desirable as it assists with driving phosphorus removal in the stripper.

The overflow from the reactor-clarifier is frequently used as elutriation water due to its low phosphorus content and the absence of nitrate or dissolved oxygen(3). Primary effluent may be a better source of elutriation water since it contains readily available organic matter to assist in the phosphorus release process. Secondary effluent is the least desirable source of elutriation water, and it should be used only if nitrification, even on a seasonal basis, does not occur in the activated sludge system.

7.3.1.2 Mainstream Processes

The mainstream nutrient removal processes include A/O, A^2/O, UCT, VIP, and Bardenpho. They involve many similar process design considerations, in spite of the variations of process unit arrangements and recycle flows. Again, the references noted earlier in the section should be consulted for detailed process design procedures for each of these systems. The reader is also referred to Chapter 3 for design considerations related to the nitrogen removal aspects of these systems. The following paragraphs discuss several design considerations related to phosphorus removal that are common to all of the mainstream systems. The primary phosphorus-related concerns are: (1) design of the anaerobic zone; (2) procedures for sludge processing; (3) capability of the process to meet effluent phosphorus limitations; and (4) selection of an appropriate SRT.

Anaerobic Zone Design. All of the mainstream systems include an anaerobic zone to stimulate the subsequent excess microbial uptake of phosphorus in an aerobic zone. The design of the anaerobic zone must provide adequate time for the phosphorus-removing microorganisms in the mixed liquor to assimilate and store the soluble organic substrate in the influent. As with the anoxic zone described in Chapter 3, the anaerobic zone is mixed but not aerated.

This zone is typically sized to provide a hydraulic residence time (HRT) between 0.9 and 2.0 hours based on process influent flow(3). For some wastewaters, an HRT at the upper end of this range results in greater phosphorus removals. This may be more important for wastewaters having a relatively low soluble BOD to phosphorus ratio, suggesting a relatively low readily biodegradable organic matter content. For such wastewaters, the longer reaction time would allow conversion of particulate BOD to soluble BOD through fermentation for subsequent assimilation by the phosphorus-removing microorganisms. In contrast, shorter HRTs may prove adequate for septic wastewaters containing a relatively high soluble BOD content.

Sludge Processing. The biological mechanism for phosphorus removal in mainstream biological phosphorus removal systems is the incorporation of phosphorus into the activated sludge cell mass, which is removed from the process as waste activated sludge (WAS). This sludge is generally thickened and then stabilized in some manner prior to ultimate disposal. Since the sludge is removed from the phosphorus release/uptake cycle subsequent to uptake, the potential for phosphorus release exists if the sludge is subjected to anaerobic conditions. If this release occurs in a process that has a significant recycle flow to the plant flow stream (such as gravity thickening), some but not all of the phosphorus removed from the system may be returned to the system in the recycle stream. This can be completely avoided only by the use of sludge processing systems that do not include anaerobic conditions, or by chemical treatment of the phosphorus-laden recycle stream prior to its return to the liquid process.

WAS at plants practicing mainstream biological nutrient removal is typically thickened using a process such as dissolved air flotation thickening to avoid anaerobic conditions and phosphorus release. Stabilization of the sludge through aerobic digestion will result in phosphorus release as cells are lysed. Likewise, anaerobic digestion will result in the release of much of the phosphorus into solution, although a portion may precipitate as magnesium ammonium phosphate, or struvite. Struvite precipitation is discussed in Chapter 4. However, the effect of this release on plant effluent quality will vary from plant to plant. For example, a study in Pontiac, Michigan(7), did not find that phosphorus present in digester supernatant from an A/O plant had a significant impact on plant effluent quality. This contrasts with the results of a similar study at the York River wastewater treatment plant where the phosphorus content of the recycle streams was quite significant(8). Regardless, if neither supernatant nor overflow were returned to the plant flow stream, anaerobic digestion could be a viable sludge stabilization process for phosphorus removal plants.

Some plants with mainstream phosphorus removal do not aerobically or anaerobically digest their sludge, but rather discharge the sludge directly to sludge drying beds, to a composting facility, or to incineration(3), thereby eliminating altogether recycle streams from sludge processes.

Effluent Phosphorus Reduction Capability. The biological phosphorus removal processes described in this chapter are all capable of substantial reductions in effluent phosphorus concentrations if operated under the proper conditions. However, to allow compliance with a specified effluent discharge standard, it may be necessary to incorporate chemical polishing of the effluent and/or effluent filtration to remove a portion of the remaining phosphorus.

A methodology for predicting the phosphorus removal capabilities of a biological system is provided in the EPA **Design Manual for Phosphorus Removal**(3). Basically, the removal of phosphorus is determined based on the net biological solids yield and an assumed fractional content of phosphorus in the biological solids. The yield is dependent primarily on the wastewater temperature, process SRT, and whether or not primary treatment precedes the secondary process. If nitrification/denitrification is practiced at the plant, the reduction of available BOD by denitrification must also be included in the phosphorus reduction calculations. If these calculations indicate an inability or marginal capability to meet the effluent phosphorus limitation, the inclusion of effluent filtration and/or chemical treatment in the plant process train will probably be necessary.

SRT Selection. The primary factor affecting the selection of the design SRT is the degree of nitrogen removal necessary. As discussed in Chapter 3, some biological nutrient removal systems are high-rate systems (low SRT), such as the A/O process, and others are low-rate systems (high SRT), such as the Five-Stage Bardenpho process. Theory and experience indicate that a strong relationship exists between the system SRT and the phosphorus removal efficiency per unit of BOD. This can become an important factor for wastewaters with a $TBOD_5/TP$ ratio less than 20:1. Therefore, mainstream biological phosphorus removal processes should generally be operated at the minimum SRT compatible with the overall treatment needs. Since both nitrification and nitrogen removal require the use of longer process SRTs, incorporation of these capabilities adversely affects the phosphorus removal capability of the process.

7.3.1.3 General Considerations

In addition to the specific design considerations discussed above, there are several additional considerations that affect the design of both sidestream and mainstream processes.

Effluent Suspended Solids. All secondary treatment processes leave a certain quantity of biological solids in the clarified effluent. Effluent discharge permits usually limit the effluent TSS concentration to a value between 5 and 30 mg/L. These solids always contain some phosphorus, which is included in the total phosphorus (TP) content of the effluent. However, in biological phosphorus removal systems the phosphorus content of the effluent TSS can become a significant fraction of the TP discharged. For Phostrip plants, the phosphorus content of the TSS is generally between 2 and 3 percent, while it can be 6 percent or more for a mainstream system(3).

To demonstrate the impact of this factor on plant effluent TP concentration, assume a TP limit of 1.0 mg P/L and a soluble phosphorus content of 0.5 mg P/L. If a mainstream process is used, with a phosphorus content of 5 percent in the biological solids, the maximum allowable TSS concentration in the effluent is only 10 mg/L to comply with the effluent phosphorus limit. This emphasizes the importance of assessing the necessary plant performance in terms of TSS reduction to meet the effluent TP requirements, and the potential need for conservative clarifier design and/or effluent filtration. Figure 7-9 indicates the effect of effluent TSS on effluent total phosphorus for solids with a variety of phosphorus contents.

Wastewater Temperature. As discussed in Chapter 3, wastewater temperature plays a major role in nitrification/denitrification process design. In contrast, temperature does not appear to play as significant a role in biological phosphorus removal. Limited studies of operating systems with wastewater temperatures as low as 5°C have not demonstrated a correlation between wastewater temperature and biological phosphorus removal(3). If anything, an improvement in the phosphorus removal process at extremely low temperatures has been observed, possibly due to a shift in microbial population to one with a higher cell yield. These results suggest that wastewater temperature may not significantly affect biological phosphorus removal process design.

Dissolved Oxygen Level in Aerobic Zone. Specific research addressing the affect of the dissolved oxygen(DO) level in the aerobic (phosphorus uptake) zone has not been reported. The mechanism of biological phosphorus uptake suggests that a higher DO level may increase the rate, but not the magnitude of phosphorus uptake(3). It has been suggested that the optimum DO concentration is 1.5 to 3.0 mg/L. If the DO is less than this, phosphorus removal may be reduced and nitrification inhibited. Bulking sludge may also develop. On the other hand, if the DO is too high, denitrification may be reduced due to excessive DO in the RAS and mixed liquor recycle streams. The corresponding increase in nitrate could likewise adversely affect the anaerobic zone operation for phosphorus removal(9).

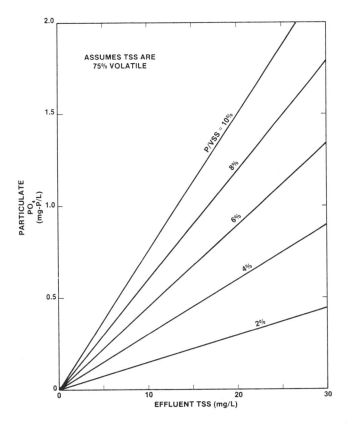

Figure 7-9. Effect of TSS on effluent phosphate.

181

7.3.1.4 Fermenter Design

As discussed above, primary sludge fermentation can be used to generate VFAs to enhance the performance of biological phosphorus removal systems. Design concepts for fermentation systems are in an evolutionary state. However, fermenters are generally designed to achieve the following objectives: (1) generate VFAs by acid phase digestion of primary sludge solids, (2) avoid hydrogen sulfide and/or methane production, since these biochemical transformations result in consumption of the VFAs generated in (1) above, and (3) elutriate the generated VFAs into the process effluent. Objectives 1 and 2 are generally achieved by designing the process for a solids residence time of approximately three days. In some cases, at least two units are provided to allow discharge of one unit, if a sulfur reducing or methanogenic population begins to develop. The third objective is achieved by recirculation of settled sludge to allow VFAs to be "washed" into the process effluent. Process schematics for primary sludge fermentation systems are presented in Chapter 6.

7.3.2 Facility Design

Once the process design has been completed, the physical facilities required to implement biological phosphorus removal must be designed. Considerations relative to design of the physical facilities for biological phosphorus removal may again be separated into the two major categories of sidestream and mainstream processes.

7.3.2.1 Sidestream Processes (Phostrip)

As described above, the facilities required to implement the Phostrip process are used to treat a portion of the RAS flow. This patented process is marketed by Biospherics, Inc., and the system is generally provided as a package designed by Biospherics. The primary components which must be considered for the system are as follows:

- o RAS feed to the stripper
- o Stripper
- o Reactor-clarifier
- o Stripped RAS return to aeration basin
- o Lime feed system

In most activated sludge treatment systems, the RAS is removed from the secondary clarifiers and pumped back to the upstream end of the aeration basin. This is generally a continuously pumped flow stream. The RAS feed to the Phostrip process may either be diverted from the pumped RAS line through a flow control valve, or a separate pumping system may be utilized drawing from a common RAS wet well. Since the total RAS withdrawal rate from the secondary clarifiers influences the operation and performance of the clarifiers, it is preferable to divert flow from the pumped RAS line to the Phostrip process to allow better control over clarifier operation. The use of a separate pumping system increases the complexity of controlling the RAS removal rate from the clarifiers. Flow metering should be provided on the RAS diversion line to allow control and monitoring of the flow. The RAS should enter the stripper basin below the water surface. A free, plunging discharge into the basin should be avoided as it will entrain air and inhibit the anaerobic processes occurring in the stripper.

The sizing of the stripper and reactor-clarifier were discussed earlier in this chapter. These components are typically supplied as part of the Phostrip package, either using prepackaged metal tankage, cast-in-place concrete basins, or existing available tankage volume. The configuration and equipment selection for these components are generally determined as a result of evaluations and negotiations between the owner, engineer, and system vendor.

The stripped sludge is removed from the stripper tank and pumped back to the RAS line leading to the aeration basins. A low-head pumping system may be required for this service. Since the stripped sludge is similar in characteristics to the RAS removed from the secondary clarifiers, similar pumping equipment may be used (e.g., nonclog centrifugal sewage pumps). This pumping system should be of the variable speed type with in-line flow measurement to allow control of the SRT in the clarifiers. Alternatively, if the stripper is designed to operate at a sufficient hydraulic gradient with respect to the main plant flow stream, it is possible that gravity return of the stripped sludge to the process may be possible. Alternative pumping and process hydraulic gradients should be considered during detailed design. Sludge density monitoring (using an in-line metering device) may be included, along with a low-density pump shutdown function.

The other flow streams associated with the stripper are the elutriation water feed and the supernatant. As discussed earlier, the elutriation water may be any of the following sources:

o   Primary clarifier effluent
o   Secondary clarifier effluent
o   Reactor clarifier overflow

If either primary or secondary clarifier effluent is used, it will typically be necessary to pump the elutriation water to the stripper tank. The elutriation water feed rate is often approximately one-half of the sludge feed rate, although the exact flow rate may not be critical to the elutriation process. This flow rate, along with the chemical nature of the water, determines the lime dosage requirements and the sizing of the reactor clarifier. This suggests the use of a constant speed pumping system to simplify design and operation of the lime feed system and reactor-clarifier.

If overflow from the reactor-clarifier is used as elutriation water, it must be pumped to the stripper tank. Detailed flow balances must be considered in the design of this system. In general, the reactor-clarifier overflow will exceed the elutriation recycle pumping rate. Consequently, it will be necessary to provide an overflow mechanism in the elutriation water pumping system and an auxiliary pumping system to handle the excess overflow. As with the other sources of elutriation water, a constant speed pumping system should be considered.

The lime feed system may be selected and provided by the Phostrip manufacturer, or it may be procured separately. The key design consideration for this system is that of providing redundancy. Lime feed systems are known for having frequent maintenance problems due to the nature of the material being handled. The most significant problem is scaling, which primarily affects the piping systems. These systems should be easily dismantled for cleaning, or a parallel standby feed system (or at least parallel piping systems) should be required. As an alternative to the use of a reactor-clarifier for lime treatment of the stripper supernatant, the supernatant may instead pass through a flash mixing compartment to which the lime is added, and then be discharged to the primary clarifiers to be removed with the primary sludge. This, of course, does not allow the separate processing and disposal of the lime sludge, and the impact on the primary sludge processing system should be considered.

7.3.2.2 Mainstream Processes

The mainstream biological phosphorus removal systems are similar in most respects to the biological nitrogen removal systems described in Chapter 3. As such, the reader is referred to that chapter for a discussion of the facility design considerations for the aerobic zones, secondary clarifiers, and anoxic zones (for processes such as A^2/O or Bardenpho which include nitrogen removal). In addition, the discussion of recycle pumping systems in Chapter 3 is equally applicable to the various recycle pumping requirements of the biological phosphorus removal process.

The primary additional component in all of the biological phosphorus removal systems is the anaerobic zone. However, even this component is virtually identical to an anoxic zone in terms of facilities design, since both include mixing without aeration. The mixing energy input should be similar to that for an anoxic basin, or approximately 50 horsepower per million gallons (MG) of volume. The same choices of mixers (propeller type or submerged turbine) are also appropriate for the anaerobic zone. An important consideration in the design of the anaerobic zone is in the discharge of the influent and recycle flows. These discharge points should always be submerged to avoid entraining air into the basin contents.

Other considerations in the design of the anaerobic zone are the detention time (volume), and whether to provide a single basin or multiple tanks in series. Tanks in series may provide improved phosphorus uptake due to the first-order reaction kinetics of the process, which benefits from a higher BOD concentration in the first stage and a higher corresponding reaction rate(5). Offsetting this benefit, however, are the increased construction costs for multiple basins over a single basin.

7.3.3 Facilities Costs

As with facility design, the facility cost considerations for biological phosphorus removal systems may be separated into the two categories of sidestream(Phostrip) and mainstream processes. The Phostrip process is a proprietary process, and the equipment package is typically supplied by the manufacturer, Biospherics, Inc. If this system is being considered for a particular application, the manufacturer should be contacted for either an estimate or a detailed price, as necessary.

The mainstream biological phosphorus removal processes include the same basic components as the biological nutrient removal systems discussed in Chapter 3. Cost guidelines for item such as concrete basins, baffle walls, mixers, aeration systems, and recycle pumping systems were provided in that chapter.

7.4 System Operation

7.4.1 Operational Considerations

Operational considerations for biological nitrogen removal are discussed in Chapter 3. Since these processes are essentially controlled in the same manner when combined with biological phosphorus removal, the reader is referred to Chapter 3 for the operation of the biological nitrogen removal portion of a combined nutrient removal system. The following discussion is limited to the phosphorus removal processes and those operational aspects of the nitrogen and phosphorus removal combined systems that result from the phosphorus removal requirements.

7.4.1.1 Unique Phostrip Considerations

The Phostrip process, unlike the mainstream processes, is essentially an additional and separate unit process that includes its own operational and control considerations. The stripper and reactor-clarifier sizes are fixed by the design. The sidestream sludge feed rate, the stripper underflow rate, and the elutriation water flow rate are the primary operational parameters. The sidestream sludge flow rate is typically constant, at a percentage (typically 10 to 30 percent) of the plant flow rate, and the elutriation water flow rate is likewise related to the sludge feed rate (typically 50 percent). These rates should be adjusted seasonally to correspond to the average plant flow rate at the time.

The primary operational control parameter for the Phostrip process is the solids residence time (SRT) in the stripper. Although the maximum SRT is fixed by stripper size, a relatively wide range of operations may be achieved by varying the underflow rate and sludge blanket level. As noted earlier, the normal range of SRT is 5 to 20 hours. The necessary SRT for a particular plant is dependent on the active biomass in the RAS and the source of elutriation water. If primary effluent is used, the required SRT may be lower as a result of the soluble BOD available.

The Phostrip process removes phosphorus both through a sidestream biological/chemical process and through enhanced uptake into the waste activated sludge (WAS). If the phosphorus effluent limitation is less stringent during a portion of the year, it may be possible to operate the system to utilize only the removal of phosphorus in the WAS. To do this, the sludge should still be sent through the stripper to provide an anaerobic contact zone to activate the sludge for excess phosphorus uptake. However, the supernatant may be returned to the plant flow stream directly and not subjected to lime treatment. This results in cost savings due to reduced chemical costs and sludge production.

An additional operational concern with the Phostrip process is the development of scale in the reactor-clarifier and the stripper (if overflow elutriation is practiced). The removal of this scale typically requires temporary shutdown of the system and the use of an acid wash.

7.4.1.2 Other Operational Considerations

The following paragraphs discuss operational considerations that are either specific to the mainstream process, or are applicable to all of the phosphorus removal processes.

TBOD$_5$/TP Ratio. As discussed earlier, the ratio of total BOD to total phosphorus (TBOD$_5$/TP) of the applied wastewater has a major impact on the performance of mainstream biological phosphorus removal systems. Generally, a TBOD$_5$/TP ratio greater than 20:1 is necessary to achieve low effluent phosphorus concentrations(3). This factor should be considered in the design of the facilities, of course, and a process that can accommodate a low TBOD$_5$/TP ratio (Phostrip) should be selected if the low TBOD$_5$/TP condition is known at the time of design. However, even a properly designed plant may experience problems with a low TBOD$_5$/TP ratio as a result of storm flows (in a combined system), lower than anticipated initial flows and loads, or seasonal flows and loading reductions.

At least three approaches are available for increasing the TBOD$_5$/TP ratio at a plant(5). The most simple approach is to remove primary clarifiers from service to overload the remaining clarifiers. This can result in a reduced performance of the clarifiers and an increase in the TBOD$_5$ carryover to the secondary process. A second approach, which may require a temporary piping modification at the plant, is to feed primary sludge directly to the secondary system. A third approach is to implement a primary sludge fermentation system to generate volatile fatty acids to feed t• the secondary process. These latter two approaches have proven effective in full-scale facilities(10).

Sludge Processing and Handling. The secondary sludge from any of the biological phosphorus removal processes is rich in phosphorus. Phosphorus tends to remain in the sludge as long as it remains in an aerobic state and aerobic digestion of the sludge does not occur. However, phosphorus will be quickly released into solution if the sludge is subjected to anaerobic conditions. This factor must be considered in the operation of the sludge handling systems. The sludge must not be allowed to remain in the secondary clarifiers for an excessive period of time, which can be accomplished by maintaining a relatively low sludge blanket. It must also not be allowed to stand for an excessive period of time in unaerated wet wells or holding basins. If the sludge is aerobically digested, the supernatant must be monitored and chemically treated, if necessary, to avoid a buildup of phosphorus in the activated sludge inventory. Anaerobic digestion and dewatering can result in a phosphorus-rich recycle stream which must be considered.

Anaerobic Zone Hydraulic Residence Time. The hydraulic residence time in the anaerobic zone is fixed by the designer in terms of volume versus the plant and recycle flow rates. If inadequate phosphorus removal is occurring at a plant which otherwise has an adequate $TBOD_5/TP$ ratio (greater than 20:1), it is possible that the anaerobic zone hydraulic residence time is inadequate. The retention time of this zone may be increased by reducing the RAS flow rate (A^2/O, Bardenpho) or the anoxic recycle rate (UCT, VIP). However, the effect of RAS flow rate reduction on secondary clarifier operation must be considered. Another approach is to convert the first anoxic zone to an anaerobic zone, while providing an anoxic region at the upstream end of the aerobic zone. With this approach, an adequate aerobic zone must be retained to ensure phosphorus uptake.

Nitrate Recycle Control. Some biological phosphorus removal processes recycle RAS to the anaerobic zone (A^2/O, Bardenpho). Other processes recycle mixed liquor from the anoxic zone to the anaerobic zone. In either case, particularly for a plant with a low $TBOD_5/TP$ ratio or an inadequate anaerobic zone retention time, the nitrate content of the recycle stream should be kept as low as possible. For RAS recycle, the only feasible alternative is to reduce the RAS pumping rate since the RAS, by virtue of the nitrification process in the aerobic zone, will contain nitrate. This must be balanced against the need to avoid anaerobic conditions in the sludge blanket. For anoxic mixed liquor recycle, the nitrate content can be reduced through careful operation of the anoxic zone to achieve complete denitrification. Nitrified mixed liquor can be decreased to reduce the nitrate loading on the anoxic zone.

Activated Sludge System SRT. For the mainstream processes phosphorus is removed through uptake in the activated sludge. This also accounts for a significant portion of the phosphorus removed in the Phostrip process. As the plant SRT is reduced, the biomass becomes more active, resulting in a greater rate of phosphorus uptake and quantity of sludge removed, or wasted, from the system each day. Consequently, an activated sludge system with biological phosphorus removal should be operated at the lowest SRT compatible with the nitrification/denitrification needs of the plant.

Process Monitoring. Biological nutrient removal plants require a greater degree of process monitoring than a typical activated sludge plant. Ideally, the following parameters should be monitored for process control from 24-hour composite samples, or at least from several grab samples taken on a daily basis:

o	Raw Sewage or Primary Effluent (unfiltered):	BOD; TKN; NH_3-N; Total P; O-PO4
o	Anaerobic Zone (filtered):	NO_3-N; O-PO_4
o	First Anoxic Zone (filtered):	NO_3-N; O-PO_4
o	First Aerobic Zone (filtered):	NO_3-N; O-PO_4; NH_3-N
o	Second Anoxic Zone (filtered):	NO_3-N; O-PO_4
o	Second Aerobic Zone (filtered/unfiltered):	NO_3-N; O-PO_4; NH_3-N / TSS; SVI; pH
o	RAS (filtered/unfiltered):	NO_3-N; O-PO_4 / TSS
o	Final Effluent(filtered/unfiltered):	NO_3-N; O-PO_4; NH_3-N / TSS

186

The distinction between filtered and unfiltered components is made to distinguish between the soluble and total compositions of the samples. The above listing is obviously based on a five-stage Bardenpho process. For other mainstream processes a similar listing should be derived.

The process monitoring requirements for a Phostrip plant are somewhat different as a result of the different processes involved. For a Phostrip plant, the following parameters should be ideally monitored either from 24-hour composite samples or multiple grab samples:

o	Raw Sewage (unfiltered):	BOD; TKN; NH_3-N; Total P; O-PO_4
o	Aeration Basin (filtered):	NO_3-N; O-PO_4; NH_3-N
o	RAS (filtered/unfiltered):	NO_3-N; O-PO_4 / TSS
o	Elutriant Water:	alkalinity
o	Stripper Underflow (filtered/unfiltered):	O-PO_4 / TSS
o	Reactor-Clarifier Overflow:	O-PO_4; pH(grab); TSS
o	Plant Effluent (unfiltered):	Total P; O-PO_4; NH_3-N; TSS; NO_3-N

7.4.2 Operational Cost Considerations

7.4.2.1 Sidestream Processes (Phostrip)

The Phostrip process has different operational cost considerations than the mainstream processes. Energy consumption for the Phostrip process includes the low horsepower drive units for the stripper mechanism and the reactor-clarifier, along with the pumping of stripper underflow sludge, lime sludge, elutriation water, and reactor-clarifier overflow. However, the primary operational cost consideration for the Phostrip process is the lime usage for stripper supernatant treatment. The lime dosage is typically 100-300 mg/L depending on the flow and alkalinity of the stripper supernatant. Lime prices vary throughout the country, with shipping being a significant additional cost depending on plant location. Quicklime (CaO) costs vary from $40 to $60 per ton in bulk, and from $70 to $95 per ton in 80-lb bags. Hydrated lime (Ca[OH]$_2$) varies from $45 to $70 per ton in bulk, and from $65 to $90 per ton in 50-lb bags. Shipping can add from $10 to $40 per ton to the above costs.

To approximate the costs for lime usage in a Phostrip process, a dosage rate must first be assumed. To be conservative, a dose of 300 mg/L should be used. The annual lime usage may then be estimated based on the dosage and elutriation water flow rate (which will approximately equal the supernatant flow rate). For example, for a plant with a design average flow of 10 MGD, an RAS feed to the stripper of 25 percent of plant flow, and an elutriation water flow of 75 percent of the stripper feed, the lime usage may be calculated as follows:

$$\text{Lime (as CaO)} = (300 \text{ mg/L}) (10 \text{ MGD}) (0.25) (0.75) (8.34) (365 \text{ day/yr})$$

$$= 1,712,306 \text{ lb/yr}$$

Assuming a delivered cost of $70 per ton for quicklime:

$$\text{Cost/year} = (1,712,306 \text{ lb/yr}) (\$70/\text{ton}) (2,000 \text{ lb/ton})$$

$$= \$60,000/\text{yr}$$

As shown by this example, the chemical costs for the Phostrip process are significant and must be included in a present worth evaluation of the process.

7.4.2.2 Mainstream Processes

The operational aspects of the mainstream biological phosphorus removal systems are essentially the same as for the mainstream biological nitrogen removal systems discussed in Chapter 3. The additional operational costs for phosphorus removal primarily include the energy costs for mixing the anaerobic zone and for pumping the anoxic recycle in some systems (UCT and VIP). Another operational cost that may be incurred with the mainstream system is that for final effluent polishing. If chemical treatment is used for polishing, the cost of the chemical will add to the system operational costs. These costs are discussed in Chapter 5 of this manual. On the other hand, if effluent filtration is used, the additional operational costs will include pumping of the backwash wastewater and periodic addition or replacement of the filter media.

7.5 Full-Scale Experience

7.5.1 General

Prior to the mid-1970s, the only phosphorus removal processes in use were the chemical removal processes, as described in Chapter 5 of this manual. In the early 1970s, the Phostrip process was introduced, with the Modified Bardenpho process following shortly thereafter. Since that time several plants using both of those processes have been placed into service, as well as other plants using subsequently developed processes such as A^2/O, UCT, and VIP. These biological processes have offered alternatives to the higher operational costs of the chemical phosphorus removal processes, with relatively little additional operational costs and moderate additional capital costs over and above that for activated sludge treatment.

7.5.2 Case Studies

This section includes descriptions of several actual operating full-scale biological phosphorus removal plants and pilot-scale studies. These examples include some of the processes discussed in this chapter, as follows:

Process Type	Plant
Phostrip	Tahoe-Truckee WWTP, California Reno-Sparks WWTF, Nevada
Bardenpho	Palmetto WWTP, Florida
A^2/O	Largo WWTP, Florida
A/O	East Boulevard WWTP, Pontiac, Michigan
A/O, A^2/O, and VIP	York River WWTP, Virginia
VIP	Virginia Initiative Plant, Virginia

Some of the plants were also described in Chapter 3 relative to their nitrogen removal capabilities. For those plants, their descriptions will be repeated here, concentrating instead on phosphorus removal capabilities.

7.5.2.1 Tahoe-Truckee WWTP, Truckee, California

<u>Facility Description</u>. The Tahoe-Truckee Wastewater Treatment plant was originally placed on line in 1978 with a design capacity of 4.8 MGD. The plant included two-stage lime precipitation and effluent filtration to meet an effluent phosphorus limit of 0.15 mg P/l. More recently (in 1983) the plant was expanded to a capacity of 9.6 MGD and a Phostrip sidestream removal process was added(11).

As shown in Figure 7-10, the plant currently includes primary treatment, a high purity oxygen activated sludge system, and effluent filtration. Primary effluent is used as elutriant for the Phostrip process. The stripper supernatant is subjected to lime treatment and then to a two-stage recarbonation process prior to passing through the effluent filters. The revised approach to phosphorus removal was made possible due to a relaxation of the effluent total phosphorus limit from 0.15 to 0.8 mg P/L.

<u>Effluent Limits</u>. Although the Tahoe-Truckee plant discharges into a leach field, it is governed by strict effluent limitations. These limits are as follows:

Parameter	Effluent Limit (mg/L)
$TBOD_5$	5
TSS	5
TP	0.8

<u>Wastewater Characteristics</u>. The wastewater treated by the Tahoe-Truckee plant is primarily domestic, but it is hardly typical. The wastewater is much colder than at most plants, ranging from 4°C to 18°C. The influent flows and BOD are highly variable due to the fact that it serves a resort area, with diurnal and weekly flow variations of 4 to 1 not uncommon. These variations create a challenge to the plant operations staff in meeting the effluent permit limits.

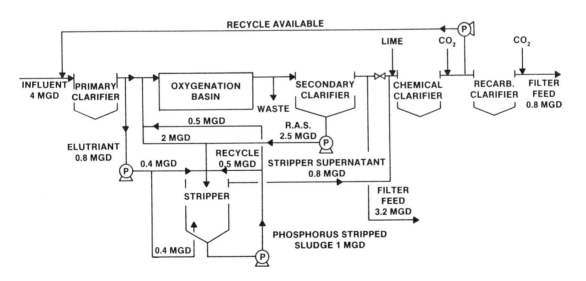

Figure 7-10. Phostrip system at Tahoe-Truckee Sanitation Agency WWTP.

189

The mean value and range of monthly average values for some of the monitored parameters for the period of July 1985 through June 1990 are listed below.

Parameter	Range	Average
Flow, MGD	2.2 - 7.1	3.6
TBOD$_5$, mg/L	98 - 206	184
TSS, mg/L	72 - 232	179
TP, mg/L	3.5 - 9.3	7.2

Operating Results. In spite of the challenge presented by the unusual wastewater characteristics and variability, the performance of the Tahoe-Truckee plant following the expansion has been excellent. Using the Phostrip process and effluent filtration, the plant has consistently met its 0.8 mg/L effluent total phosphorus limitation. The plant has realized an annual operating cost savings of $300,000 per year for chemical purchase over the previous chemical precipitation phosphorus removal process, along with a corresponding sludge production decrease of 650 tons per year. In addition, the plant has demonstrated a high degree of mechanical reliability. The effluent characteristics are presented below. A probability plot of effluent total phosphorus concentrations for this plant is presented in Figure 7-12.

Parameter	Range	Average
TSS, mg/L	1 - 4	1
TP, mg/L	0.16 - 0.58	0.35

Summary. The Tahoe-Truckee WWTP demonstrates the capabilities of the Phostrip process, when coupled with effluent polishing by filtration, to consistently treat wastewater to an effluent total phosphorus level of less than 1 mg/L. Although the Phostrip process has exhibited mechanical reliability at this plant, this has not been the case with all Phostrip plants(12). Care must be taken in the design to specify proper corrosion resistant materials and to provide redundancy in the lime feed system to ensure uninterrupted operation.

7.5.2.2 Reno-Sparks Wastewater Treatment Facility, Cities of Reno and Sparks, Nevada

Facility Description. The Reno-Sparks WWTF is discussed in Chapter 3 with regard to its nitrogen removal capabilities. This facility also provides a good example of the Phostrip process. Figure 3-11 in Chapter 3 is a flow schematic for the facility.

Effluent Limits. The Reno-Sparks WWTF discharge limits, based on a monthly average, are as follows:

Parameter	Discharge Limits
Flow, MGD	40
BOD$_5$ (inhibited), mg/L	10
BOD$_5$ (uninhibited), mg/L	20
Suspended solids, mg/L	20
Total N, mg/L	5
Total P, mg/L	0.4[a]

[a]Based on a flow of 40 MGD; mass limitation is 134 pounds per day.

190

Wastewater Characteristics. The average influent characteristics for 1986 for the plant are listed below. The actual values experienced are still somewhat less than the design values.

Parameter	Actual	Design
Flow, MGD	26.5	40
BOD_5 (inhibited), mg/L	156	--
BOD_5 (uninhibited), mg/L	188	275
Suspended solids, mg/L	177	250
Total P, mg/L	8.5	10

Operating Results. The final effluent characteristics for 1986 (monthly averages) are highlighted in the following table. Also included in the table are more current data (July 1989-July 1990). During the July 1989-July 1990 period, the monthly average total phosphorus concentration never exceeded the 0.4 mg P/L limit; the discharge limit has not been exceeded since December 1987. For the period July 1989-July 1990 the total phosphorus concentrations ranged from 0.12 to 0.34 mg/L, based on a monthly average.

Parameter	1986 Average	July '89-July '90 Average
BOD_5 (inhibited), mg/L	5.5	--
BOD_5 (uninhibited), mg/L	10	--
Suspended solids, mg/L	7.3	--
Total P, mg/L	0.33	0.21

A probability plot of effluent total phosphorus concentrations for this plant is presented in Figure 7-12.

Summary. The Reno-Sparks WWTF has been producing effluent well within its permitted discharge limits. The Phostrip process, coupled with effluent filtration, has been operating very well, producing a very high quality effluent with a total phosphorus concentration as low as 0.12 mg P/L.

7.5.2.3 Palmetto WWTP Palmetto, Florida

Facilities Description. The Palmetto WWTP is presented in Chapter 3 as an example of a successful biological nitrogen removal process. The plant is also an excellent example of the capabilities of a properly operated Bardenpho plant to achieve consistent removals of phosphorus. Details of the facility design are presented in Chapter 3(13).

Effluent Limits. The effluent monthly average permit limits for the Palmetto plant are as follows:

Parameter	Effluent Limit
$TBOD_5$, mg/L	5
TSS, mg/L	5
Total Nitrogen, mg/L	3
TP, mg/L	1

Wastewater Characteristics. The wastewater treated by the Palmetto plant is a domestic wastewater of average strength. The wastewater characteristics observed during the period of January 1984 through November 1987 are lower than the values on which the plant design was based, as shown below:

| | Design | Observed Value | |
Parameter	Value	Average	Range
$TBOD_5$, mg/L	270	158	87 - 232
TSS, mg/L	250	135	70 - 224
Total Nitrogen, mg/L	43	33.1	15.1 - 45.9
TP, mg/L	14	5.3	0.7 - 7.9
Temperature	---	---	18°C - 25°C
$TBOD_5$/TP Ratio	19:1	30:1	---

Operating Results. During the period from January 1984 to November 1987 the plant was loaded above its hydraulic design capacity, but it was under loaded with respect to organic loading. While the plant flow ranged as high as 178 percent of design, the plant $TBOD_5$ loading was only about half of design. The plant has consistently met its effluent permit limitations for $TBOD_5$, TSS, and total nitrogen.

The plant was unable to meet its effluent phosphorus limitation using the biological process alone, although phosphorus reductions of 65 percent were observed. It is suspected that the phosphorus removal limitation was the direct result of the long sludge age (and hence low sludge production) made necessary by nitrification under the lightly loaded conditions. The addition of alum to the filter influent has been necessary to further reduce the effluent phosphorus to the permit limit. A probability plot of effluent total phosphorus concentrations for this plant is presented below in Figure 7-12 at the end of the chapter.

Summary. The Palmetto plant demonstrates the capability of the Bardenpho process to achieve combined reductions of nitrogen and phosphorus. However, it also indicates the potential conflict between the nitrification process needs and the phosphorus removal needs under certain conditions. Although the Palmetto plant could not achieve its effluent standard for phosphorus using the biological process alone, the chemical requirements for phosphorus removal were reduced significantly due to the removal of phosphorus in the secondary treatment system.

7.5.2.4 Largo WWTP, Largo, Florida

Facility Description. The Largo WWTP is described in Chapter 3 with respect to its nitrogen removal capabilities. Using the A^2/O process, the Largo plant also has the capabilities to remove phosphorus. The plant includes three parallel treatment trains to provide a total plant design capacity of 15 MGD. As shown in Figure 3-13, the plant includes preliminary treatment, primary clarification, secondary treatment, effluent filtration, and disinfection. Secondary treatment, including nitrogen and phosphorus removal, is accomplished using the A^2/O process described earlier in this chapter. The A^2/O process at the Largo plant differs from the Bardenpho process at the Palmetto plant in that it includes only a single anoxic zone and a single aerobic zone. It is also operated at a higher rate, with a typical SRT of less than 10 days. A total HRT of 4.2 hours is provided in the secondary system, with 0.8 hour in the anaerobic zone, 0.5 hour in the anoxic zone, and 2.9 hours in the aerobic zone(4).

Effluent Limits. The Largo plant is subjected to effluent limitations of 5 mg/L each for $TBOD_5$ and TSS. The total phosphorus limitations at the plant are 4 mg/L, 6 mg/L, and 9 mg/L on an annual average, monthly and weekly basis, respectively.

Wastewater Characteristics. The wastewater treated at the Largo WWTP is a medium strength, primarily domestic wastewater. The influent wastewater characteristics are as follows:

Parameter	Average (mg/L)	Range (mg/L)
$TBOD_5$	200	113 - 375
TSS	325	143 - 511
TKN, maximum	30	---
NH3-N, maximum	20	---
TP	9.5	5.0 - 16.8
$TBOD_5$/TP Ratio	18:1	---

The average plant flow during the period from January 1984 to November 1987 was 9.9 MGD, which is approximately 70 percent of the plant design capacity. The plant has performed within the permit limits for $TBOD_5$ and TSS, with plant effluent averages of 5 mg/L and 4 mg/L, respectively. However, the plant has achieved variable results in total phosphorus removal with monthly average effluent values varying from 0.5 to 4.6 mg/L and averaging 2.4 mg/L. Although the plant complied with its permit limitations throughout the period, it should be noted that these limits are quite lax when compared with most other phosphorus-limiting discharge permits. An improving trend occurred, however, over the subject period with effluent phosphorus values consistently less than 2.0 mg/L after September 1986. It is suspected that this improvement resulted from a change in plant operations. A probability plot of effluent total phosphorus concentrations for this plant is presented in Figure 7-12 at the end of this chapter.

Summary. The Largo WWTP demonstrates the capability of the A^2/O process to achieve at least modest reductions in effluent phosphorus. The inability of this plant to achieve reductions equal to those of the Palmetto WWTP may be the result of a lower $TBOD_5$/TP ratio, as well as differences in operation. Also, the RAS for a Bardenpho system is typically lower in nitrate content since the nitrate is largely removed in the process. The reduced nitrate loading on the anaerobic zone may be expected to improve the performance of that zone.

7.5.2.5 East Boulevard WWTP, Pontiac, Michigan

Facility Description. The East Boulevard plant includes preliminary treatment, primary clarification, and secondary treatment. The plant effluent is then transmitted to another plant where it is processed through tertiary sand filters. The secondary treatment process for this plant is comprised of four parallel activated sludge trains, each having a design capacity of 3.5 MGD. As part of an EPA-sponsored demonstration project, two of the trains were converted to the A/O process through the addition of baffle walls and mixers. At the 3.5-MGD design flow, the system has an anaerobic HRT of 1.7 hours and an aerobic HRT of 6.4 hours.

Effluent Limits. The East Boulevard plant has seasonally varying effluent limits for $TBOD_5$ and TSS, varying from 7 to 15 mg/L for $TBOD_5$ and from 20 to 30 mg/L for TSS. The plant is required to nitrify to meet a seasonally varying NH_3-N standard ranging from 3.2 to 19.3 mg N/L. An effluent limit of 1 mg P/L for total phosphorus is in effect year-round.

Wastewater Characteristics. The wastewater treated at the East Boulevard plant is an equal mixture of domestic and industrial wastewater. The following influent wastewater parameters were observed during the demonstration project:

Parameter	Average	Range
$TBOD_5$, mg/L	228	136 - 340
TSS, mg/L	213	106 - 340
TP, mg P/L	3.18	2.6 - 3.85
$TBOD_5$/TP Ratio	70:1	---
TKN, mg N/L	22.5	17.0 - 28.5
NH3-N, mg N/L	14.8	10.8 - 18.3
Temperature	---	10°C - 20°C

Operating Results. During the demonstration project, the plant was operated in a series of seven steady-state phases, with each phase lasting from 1-1/2 to 2 months. The total process HRT ranged from 4.9 hours to 9.7 hours, and the process SRT ranged from 11.9 to 24 days. The aeration basin MLSS was maintained between 2,500 and 3,000 mg/L. The effluent $TBOD_5$ ranged from 6 to 17 mg/L, with an average of 10 mg/L. The effluent TSS ranged from 6 to 11 mg/L, with an average of 8 mg/L. The effluent TP varied from 0.34 to 0.9 mg P/L, with an average of 0.59 mg P/L. This compares favorably to the effluent total phosphorus level in the non-A/O process trains, which varied from 1.5 to 2.0 mg P/L.

A probability plot of effluent total phosphorus concentrations for this plant is presented in Figure 7-12.

Summary. The East Boulevard WWTP demonstrates the potential capabilities of the biological phosphorus removal processes under favorable conditions (i.e., high $TBOD_5$/TP ratio). It is particularly impressive that an average effluent level of 0.59 mg P/L was achieved without effluent polishing.

7.5.2.6 York River WWTP, Hampton Roads Sanitation District (HRSD), Virginia

Facility Description. The York River WWTP has a design capacity of 15 MGD. The plant process train includes preliminary treatment, primary clarification, secondary treatment, and effluent disinfection.

In 1986, two of the plant's six aeration basins were converted to allow operation in either the A/O, A^2/O, or VIP modes through the addition of baffle walls and mixing equipment. This conversion, shown schematically in Figure 7-11, resulted in approximately one-half of the basin in anaerobic and anoxic zones, and the remaining one-half retained as an aerobic zone. This conversion was for the purpose of demonstrating the capabilities of a biological nutrient removal process to the Virginia State Water Control Board for potential integration into other plants in Virginia(8).

Effluent Limits. The York River WWTP does not have effluent permit limits for nutrients. The demonstration project was intended to provide information for use at other plants in the state facing current or pending effluent nutrient limitations.

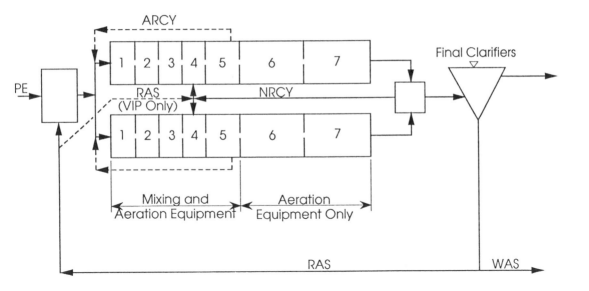

(Note: solid lines are for A/O and A²/O operation; dashed lines are for VIP operation)

Figure 7-11. York River WWTP secondary treatment process (A/O, A²/O, VIP).

Wastewater Characteristics. The wastewater received at the York River plant is primarily domestic in origin, but with significant infiltration/inflow to the collection system. During the period from December 1986 through October 1989, the following primary effluent characteristics were observed:

		Operating Mode		
	A/O			
Parameter	Period 1[a]	Period 2[b]	A²/O[c]	VIP[d]
$TBOD_5$, mg/L	117	169	206	107
TSS, mg/L	97	105	93	77
TP, mg P/L	10.5	9.6	12.8	6.8
$TBOD_5$/TP Ratio	11.0	17.2	16.1	16.0
TKN, mg N/L	28	28	25	27
NH3-N, mg N/L	21	22	20	20

[a]December 1986 through July 1987
[b]November 1987 through February 1988
[c]August 1987 through October 1987
[d]June 1988 through October 1989

<u>Operating Results</u>. The following results were obtained for the various operating conditions:

| Parameter | Operating Mode | | | |
| | A/O | | | |
	Period 1[a]	Period 2[b]	A^2/O^c	VIP[d]
Flow, MGD	8.7	7.0	5.9	8.2
$TBOD_5$, mg/L	12	10	4	10
TSS, mg/L	7	7	6	7
TP, mg P/L	3.4	3.9	4.2	1.3
TKN, mg N/L	15	7	2	9
NH3-N, mg N/L	13	6	1	8
NO_x-N, mg N/L	2	2	4	4
TN, mg N/L	17	9	6	13

[a]December 1986 through July 1987
[b]November 1987 through February 1988
[c]August 1987 through October 1987
[d]June 1988 through October 1989

During A/O Period 1 and VIP operation the wastewater was relatively weak. $TBOD_5$ and TSS concentrations were higher during the A/O Period 2 and A^2/O operation. Beginning in 1988, a ban on phosphate laundry detergents went into effect which resulted in a change in wastewater characteristics, particularly the $TBOD_5$/TP ratio. These effects were initially observed in 1987 as products were changed to allow compliance with the ban by January 1, 1988.

As indicated in the above table, wastewater flows typically exceeded the 5 MGD design value for the biological nutrient removal retrofit. Consequently, the facility was operated at relatively short HRTs. Average monthly values ranged from 2.3 to 5.9 hours. In spite of this, effluent $TBOD_5$ and TSS concentrations were routinely below the 30 mg/L secondary treatment limits for these parameters. The facility was generally operated to preclude nitrification in the A/O mode and to cause nitrification in the A^2/O mode; both nitrifying and non-nitrifying periods were experienced with the VIP mode. Monthly average effluent nitrogen concentrations in the 5 to 10 mg N/L range were observed when nitrification was relatively complete.

Significant removal of total phosphorus was obtained during all operating periods. An average of 6.7 mg P/L of total phosphate was removed by the biological process for the four operating periods. However, effluent total phosphate concentrations were still in the 3 to 4 mg P/L range due to the high influent total phosphate concentrations and relatively low $TBOD_5$/TP ratios. Effluent total phosphate concentrations were lower during VIP operation, partially due to the lower influent total phosphate concentration. A probability plot of effluent total phosphate concentrations is presented in Figure 7-12 for operation in the A/O and A^2/O, and the VIP modes.

<u>Summary</u>. The York River demonstration project shows that significant removal of total phosphate can be achieved by mainstream biological phosphate removal processes. However, effluent concentrations were 3 mg P/L or greater due to the high influent concentrations and the $TBOD_5$/TP ratios being less than 20:1. Improved performance was achieved during the VIP operating mode, partially due to the presence of a weaker wastewater.

7.5.2.7 Virginia Initiative Plant (VIP) Pilot Study, Hampton Roads Sanitation District (HRSD), Virginia

<u>Facility Description</u>. An extensive pilot study was conducted to provide design criteria for the expansion of the Lamberts Point WWTP(6). This expansion, currently under construction, will increase the plant capacity to 40 MGD. The plant process train includes preliminary treatment, primary treatment, secondary treatment, and effluent disinfection. The secondary treatment process includes biological nitrogen and phosphorus removal capabilities.

The nutrient removal process resulting from the pilot study is called the Virginia Initiative Plant (VIP) process. The VIP process was previously shown schematically in Figure 7-7. The VIP process includes the three typical zones of a combined biological phosphorus and nitrogen removal process: (1) anaerobic; (2) anoxic; and (3) aerobic.

The VIP process differs from other biological nutrient removal processes in that the RAS is recycled to the anoxic basin (downstream of the anaerobic basin), and denitrified mixed liquor is recycled from the anoxic zone to the anaerobic zone. The intent of this configuration is to improve performance of the anaerobic zone by greatly reducing, or even eliminating, nitrate loading on the anaerobic zone. Nitrate loading was cited earlier as an impediment to phosphorus removal at the York River WWTP.

The full-scale VIP will remove phosphorus year-round and nitrogen on a seasonal basis. The anaerobic and anoxic zones constitute 34 percent of the secondary reactor volume, with an HRT for the overall secondary process of 6.5 hours at 40 MGD.

<u>Effluent Limits</u>. The VIP effluent permit currently only limits $TBOD_5$ and TSS to 30 mg/L each, with no limitations on nutrients. However, the HRSD took the initiative of incorporating partial nutrient removal into the plant to protect the water quality of the Chesapeake Bay. A condition of the plant design was that nutrient removal must be achieved within conventional secondary treatment reactor sizing. This constraint resulted from restrictions on grant funding relative to the actual permit limitations. Therefore, the following goals were established for nutrient removal:

o   Phosphorus   67 percent removal, year-round

o   Nitrogen   70 percent removal for wastewater temperatures above 20°C, less for lower temperatures

<u>Wastewater Characteristics</u>. The wastewater treated during the pilot study, and typical of that to be treated by the full-scale VIP, is a relatively weak domestic wastewater. The wastewater characteristics observed during the pilot study are as follows:

Parameter	Average	Range
$TBOD_5$, mg/L	142	109 - 199
TSS, mg/L	133	98 - 152
Total Nitrogen, mg/L	25.0	21.2 - 29.3
TP, mg/L	5.2	4.2 - 6.4
$TBOD_5$/TP Ratio	27.3	---
Temperature	---	13°C - 25°C

197

Operating Results. During the pilot study, the system HRT ranged from 4 to 8 hours, and the SRT ranged either from 5 to 6 days or from 10 to 11 days, depending on wastewater temperature. The system MLSS concentration varied form 1,200 to 3,000 mg/L.

The pilot plant effluent $TBOD_5$ and TSS concentrations were well within the permit limitations, with average values of 8 mg/L for $TBOD_5$ and 10 mg/L for TSS. An oversized secondary clarifier was partly responsible for this excellent performance. The effluent total phosphorus averaged 1 mg/L for the entire study. The relatively high $TBOD_5$/TP ratio was partially responsible for this low effluent phosphorus level, along with the absence of sludge processing recycle flows in the pilot plant. A probability plot of effluent total phosphorus concentrations for this plant is presented in Figure 7-12.

Summary. The VIP pilot study demonstrates the capability of the VIP process to achieve low effluent phosphorus levels when treating a wastewater with an adequate $TBOD_5$/TP ratio. The elimination of nitrate recycle to the anaerobic zone is partially responsible for this excellent performance. The low effluent TSS resulting from the oversized clarifier is also partially responsible for the low phosphorus levels due to the reduction in particulate phosphorus in the effluent.

7.5.2.8 Conclusion

The case histories described above demonstrate the implementation of biological phosphorus removal processes (Phostrip, A/O) and of combined biological phosphorus/nitrogen removal processes (A^2/O, Bardenpho, VIP) in full-scale wastewater treatment. Figure 7-12 presents an overall comparison of the case histories discussed here. These case histories indicate the importance of the design and operational factors discussed in this chapter, such as the $TBOD_5$/TP ratio and nitrate recycle. It is apparent from these examples that effluent phosphorus reductions to the 1 mg P/L level or less require an adequate $TBOD_5$/TP ratio and elimination of nitrate recycle for the mainstream processes, or the use of the Phostrip process. However, these factors are not as critical, if a relatively high effluent phosphorus limit must be met, such as 3 mg P/L.

It is clear from the data presented in Figure 7-12 that a wide range of effluent total phosphorus concentrations can be produced by biological phosphorus removal facilities. Process type and wastewater $TBOD_5$/TP ratio are two factors which affect process effluent quality. One approach for assessing the phosphorus removal capability of biological phosphorus removal facilities is use of the $TBOD_5$ to phosphorus removal ratio(8). This ratio is calculated as follows:

$$\frac{TBOD_5 \text{ in Biological Process Influent}}{(TP \text{ in Biological Process Influent} - SP \text{ in Biological Process Effluent})}$$

where SP is the concentration of soluble phosphate. When the effluent SP is significant, i.e., above about 1 mg P/L, the full phosphorus removal capability of the process is being used.

Table 7-2 summarizes $TBOD_5$ to phosphorus removal ratios for several facilities, including some described above. Data from Palmetto are for a period during which alum addition was not practiced. The York River A/O and A^2/O data illustrate the impact of nitrification on the phosphorus removal capability of processes which recycle RAS directly to the anaerobic zone. The VIP pilot data illustrate that significant differences in phosphorus removal capabilities can exist between two processes. Comparison of the Fayetteville and the York River nitrifying A/O and A^2/O data suggests that similar performance may be observed for similar applications at different locations.

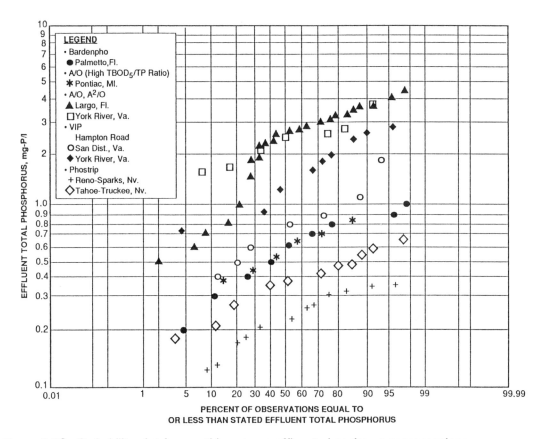

Figure 7-12. Probability plot for monthly average effluent phosphorus concentrations.

Table 7-2. Biological phosphorus removal capability.

Location	Process	TBOD$_5$ to Phosphorus Removal Ratio (mg BOD$_5$/mg TP)	Effluent Soluble P (mg P/L)	Effluent Nitrogen (mg/L)			SRT (days)	Data Period (months)	Reference
				TN	NH$_3$-N	NO$_x$-			
Palmetto, FL	Bardenpho	27.3	2.9	2.1	--	1.3	14-20	4	13
Fayetteville, AR									
Full scale	A/O	21.0	2.1	--	0.9	--	10.6	6	14
Pilot scale	A/O	22.8	2.0	9.7	1.1	7.3	4.6	3	15
York River, VA A^2/O, A/O									
Nitrifying	A^2/O, A/O	23.1	3.8	6.8	0.5	5.0	9.6	4	8
Non-nitrifying	A/O	14.0	2.7	18.5	16.4	0.8	5.9	6	8
VIP Process	VIP	<17.6	0.7[a]	8.1	1.5	4.9	7.2	9	8
VIP Pilot, VA									
VIP	VIP	10.5	1.6	8.2	0.1	6.9	6.0	--	6
A^2/O	A^2/O	18.6	5.8	7.6	0.8	5.2	5.0	--	6

[a]suggests that not all of the phosphorus removal capability of the process was used.

199

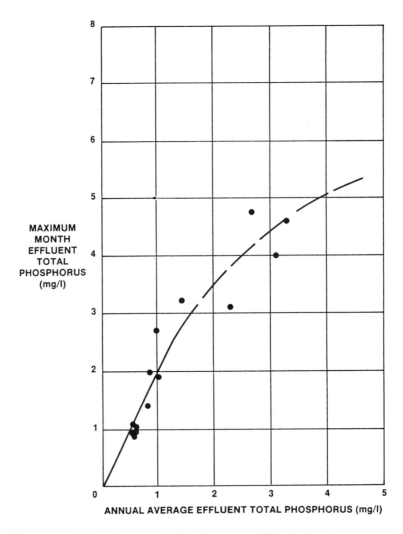

Figure 7-13. Effluent total phosphorus variability for seven biological nutrient removal facilities.

Figure 7-13 further indicates the variability in performance observed for a number of biological phosphorus removal facilities, including several of those discussed above(4). The highest monthly average effluent total phosphorus concentration for a given year is plotted as a function of the overall annual average effluent total phosphorus concentration.

The results indicate that the monthly maximum is about double the long-term average effluent quality for average effluent total phosphorus concentrations of 1.5 mg/L or less, dropping off to 50 percent greater at an average effluent concentration of 3.0 mg/L. This ratio is greater than that observed for nitrogen removal as shown in Chapter 3, but it is typical of that for other pollutants such as $TBOD_5$, TSS, and NH_3-N at a plant producing a high quality effluent. This variability must be anticipated in the selection and design of a process for phosphorus removal.

7.6 References

1. Hong, S. N., *et al*. A biological wastewater treatment system for nutrient removal. Presented at the 54th Annual WPCF Conference, Detroit, Michigan, October 4 - 9, 1981.

2. Levin, G. V., and J. Shapiro. Metabolic uptake of phosphorus by wastewater organisms. **Jour. Water Pollut. Control Fed.**, **37**, 800, 1965.

3. Environmental Protection Agency. **Design Manual: Phosphorus Removal**. 1987.

4. CH2M HILL. **Biological Nutrient Removal Study**. Presented to the Virginia State Water Control Board, 1988.

5. Ekama, G. A., G. v. R. Marais, and I. P. Siebritz. Biological excess phosphorus removal. In **Theory, Design, and Operation of Nutrient Removal Activated Sludge Processes**. Prepared for the Water Research Commission, Pretoria, South Africa, 1984.

6. Daigger, G. T., G. D. Waltrip, E. D. Romm, and L. M. Morales. Enhanced secondary treatment incorporating biological nutrient removal. **Jour. Water Pollut. Control Fed.**, **60**, 1833, 1988.

7. Kang, S. J., *et al*. A year's low temperature operation in Michigan of the A/O system for nutrient removal. Presented at the 58th Annual WPCF Conference, Kansas City, Missouri, October 1985.

8. Daigger, G. T., L. M. Morales, J. R. Borberg, and G. D. Waltrip. Full-scale and pilot-scale experience with the VIP process. Presented at the First Australian Conference on Biological Nutrient Removal (BNR1), Bendigo, Australia, 1990.

9. Ekama, G. A., I. P. Siebritz, and G. v. R. Marais. Considerations in the process design of nutrient removal activated sludge processes. Selected Papers on Activated Sludge Process Research at the University of Capetown, Capetown, South Africa, April 1982.

10. Oldham, W. K. Full-scale optimization of biological phosphorus removal at Kelowna. Presented at the IAWPRC Post-Conference Seminar on Enhanced Biological Phosphorus Removal from Wastewater, 1984.

11. Svetich, R., C. Woods, and R. Prettyman. Mainstream biological phosphorus removal--a cost-effective method of retaining pristine surface waters. Presented at the 58th Annual WPCF Conference, Kansas City, Missouri, 1985.

12. Walsh, T. K., B. W. Behrman, G. W. Weil, and E. R. Jones. A review of biological phosphorus removal technology. Presented at the 56th Annual WPCF Conference, Atlanta, Georgia, 1983.

13. Burdick, C. R, D. R. Refling, and H. D. Stensel. Advanced biological treatment to achieve nutrient removal. **Jour. Water Pollut. Control Fed.**, **54**, 1078, 1982.

14. CH2M HILL. **Fayetteville Pilot Plant Study Final Report**. Prepared for the City of Fayetteville, Arkansas, 1986.

15. City of Fayetteville, Arkansas. Operating records. 1988.

Chapter 8

Case Studies in Biological Phosphorus Removal

8.1 Introduction

The first observations of biological phosphate removal were made in high rate activated sludge plants designed and operated to limit nitrification. Such plants as the Rilling Road plant in San Antonio and the Back River plant in Baltimore produced excellent results(1), even though the operators were not aware of the mechanisms and as such were not able to optimize operations for phosphorus removal. In 1974, Barnard(2) suggested that it was necessary for the sludge or mixed liquor of an activated sludge plant to pass through an anaerobic phase absent of dissolved oxygen or nitrate and then through an aerated phase in order to activate the phosphorus removal mechanism. This theory was derived from the observation by Milbury(1) that all plants that had high removals of phosphate tended to show an initial release of phosphate near the inlet end of the activated sludge process.

Much of what is known today about biological phosphorus removal has been learned through experiences in full-scale operations. Therefore, it is necessary to study the behavior of full-scale plants to fully understand biological removal systems. It will be pointed out later that it is almost impossible to simulate full-scale behavior in the laboratory and while models may be developed from laboratory studies, they need to be confirmed using data from full-scale operations. Fortunately, it appears easier to obtain desired performance in the field than in the laboratory.

When Barnard first proposed the use of an anaerobic basin in 1974, full-scale experimentation soon followed and the construction of a number of plants started in that year and early in 1975. The Goudkoppies plant was already under construction as a denitrification plant in late 1974 when it was decided to add an anaerobic basin. Since nitrification was already a requirement, most plants designed at this time were of the three-stage or five-stage Bardenpho type. Only the Waterval plant for the City of Germiston, also started in 1974, was a non-nitrifying anaerobic/aerobic plant. The flow diagram for this plant is similar to the A/O process patented in April 1976.

During this initial period of construction, the mechanism of biological phosphorus removal was still a mystery. Fuhs and Chen(3) proposed in 1975 that the removal of phosphates in the Phostrip process was mainly due to the selective growth of **Acinetobactor**. These obligate aerobic bacteria use acetates as feedstock and accumulate a huge surplus of phosphorus which is stored as polyphosphate under aerobic conditions and is released to the surrounding liquid as orthophosphate during anaerobiosis.

Nicholls and Osborn(4) proposed that the role of the anaerobic stage was to allow for the production of short chain volatile fatty acids(VFAs), such as acetates, through the fermentation of the incoming organic feed as well as for the uptake of acetates by the **Acinetobacter** in the anaerobic zone. They proposed that **Acinetobacter** derived energy from the phosphorus pool in their cells and that they used this energy for the transport of acetate across the membrane of their cells under anaerobic conditions where it could be stored in the form of poly-ß-hydroxybutyrate (PHB) which would be further metabolized when these obligate aerobic organisms pass through the aerobic stage. Since acetate is the end product of fermentation and since no other organisms, with the exception of methane bacteria, could utilize the acetates in the absence of an electron acceptor, this ability gives these organisms a selective advantage and allows them to grow preferentially. It was apparent that nitrate entering the anaerobic stage could serve as electron acceptors for the growth of other organisms which then would consume short chain fatty acids to the detriment of the **Acinetobactor**.

The theory of Nicholls and Osborn reconciled most of the observations concerning biological phosphorus removal. In the full-scale plug flow plants, fatty acids were produced in the sewers and force mains leading to the plants and the contact period was provided in under aerated zones at the head of the aeration basin. These zones were caused by the rolling action of the mixed liquor, the high oxygen demand and very inefficient aeration. In the Phostrip plant, fatty acids were generated by fermentation of the activated sludge itself while contact between the organisms and the VFAs took place in the gravity thickener. Since the influent wastewater was fed directly to the aeration basin the fatty acids in the influent wastewater were not used in the biological phosphorus removal process.

8.1.1 Primary and Secondary Releases of Phosphorus

It appeared from the very first observations of biological phosphorus removal that a release of phosphorus preceded the uptake. Milbury(1) noted that such a release took place in all the U.S.A. plants that removed phosphorus. As the role of VFAs became evident, researchers demonstrated the relationship between acetate addition and phosphorus release, as shown in Figure 8-1. The apparent problem was that the uptake behavior upon aeration was inconsistent.

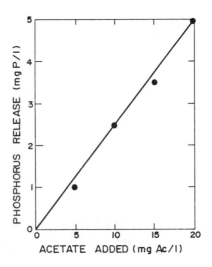

Figure 8-1. Phosphorus release with acetate addition.

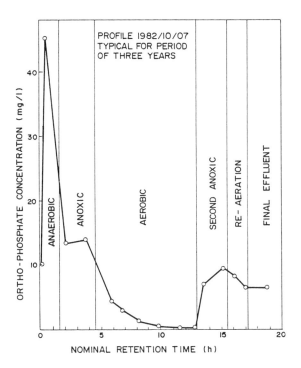

Figure 8-2. Phosphorus profile through Randfontein plant.

In the Phostrip process the released phosphorus was removed from the system and treated with lime, since all the released phosphorus could not be taken up again. An anomaly was also observed in a Bardenpho plant in which phosphorus was completely taken up by the bacterial mass by the end of the aeration basin and was then again released in the second anoxic basin. A typical example of a profile through this plant is shown in Figure 8-2. Due to favorable conditions for denitrification in the aeration basin no nitrates were fed to the second anoxic basin. Phosphorus was released, but could not be taken up again by aeration. Similar observations in other plants led to the conclusion that release of phosphorus may not always be associated with acetate uptake and that when such conditions prevail, no phosphorus uptake will be possible upon reaeration since the energy needed for uptake is not available(5).

Fuhs and Chen(3) bubbled CO_2 through an activated sludge sample taken from a phosphorus removal plant. Substantial release of phosphorus took place. In the context of stripping of phosphorus, this may have been desirable, but it is clear that such "stripped" phosphorus would not be taken up upon reaeration since no energy in the form of VFAs was available to the organisms.

Barnard(5) referred to a release associated with intake of VFAs as "primary release" and that caused by anaerobiosis in the absence of VFAs as "secondary release". It follows that both types of release could take place at the same time. Thus, while it is possible to generate acetates in the anaerobic zone of the Bardenpho plant, leading to acetate uptake by the organisms and primary release, secondary release will also take place. It is surmised that in the Phostrip process some of the phosphorus must be released by the acetate generated in the stripping tank, but that most of the phosphorus is released by the secondary mechanism. Insufficient energy would then be available to take up all of the phosphorus, hence the need for stripping and removal of excess phosphorus by lime treatment.

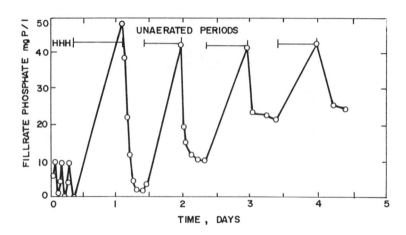

Figure 8-3. Phosphorus uptake and release.

Operations at the Bardenpho plant in Kelowna, British Columbia provide an example as to how, in practice, secondary releases can affect performance. The release of phosphorus in the anaerobic zone was doubled by switching off the stirrers and allowing the sludge to thicken in an upflow unit. However, this led to a deterioration in effluent quality. Prior to this experiment, theory suggested that the energy available through primary release was just sufficient to take up all released phosphorus. In this instance, the additional phosphorus released by switching off the stirrers could not be taken up and, therefore, it appeared in the effluent. It appears that about 4 mg/L of acetates as COD are required for the uptake of 1 mg/L of phosphorus. If sufficient acetates are available, some or all of the secondary-released phosphorus may be removed.

Earlier findings by Wells(6), illustrated in Figure 8-3, also could be interpreted in terms of primary and secondary release. If one assumes that the sludge used had a large population of phosphorus-removing organisms and the sludge contained a reservoir of organic carbon, since it was derived from a high rate plant, phosphorus released during anaerobic periods could then be taken up again during aeration. As the reservoir of food is diminished, less primary release and more secondary release takes place, resulting in slower uptake rates and incomplete uptake.

8.1.2 Role of Different Short-Chain Carbohydrates

Nitrate interferes with the biological removal of phosphorus in two ways. First, the presence of nitrates in the anaerobic zone prevents fermentation, since the organisms could derive more energy by anaerobic respiration using nitrates as an electron acceptor. Thus no fatty acids would be produced. Second, even when such fatty acids are present in the influent, nitrates could serve as an electron acceptor for heterotrophic organisms using acetate as feed. Thus little of the acetates would be available to the organisms that need them for biological phosphorus removal. However, when short-chain fatty acids of the type required by these organisms are present in the influent in large enough quantities, good phosphorus removal may be possible even in the presence of nitrates.

The role of short chain carbohydrates was best demonstrated by Gerber et al.(7) who fed nitrates together with various short chain carbohydrates to sludge from a biological phosphorus removal plant. The results are shown in Figure 8-4. Note that only when acetic acid, propionic acid and formic acid were present in the feed did phosphorus release start before the nitrates were completely reduced.

206

Note also how the rate of release declines from acetate to propionate to formate. It can be concluded that these are the only VFAs that could be used directly, with optimal results occurring with acetates. Butyric acid produced no release of phosphorus until all the nitrates were reduced. At this point some fermentation took place to reduce the butyric acid to acetic acid resulting in a high rate of release until the butyric acid was consumed. Interestingly, release continued after the substrates were consumed, but at a lower rate. This could represent secondary release. Even with butyric acid one can detect a lowering of the rate of release after consumption of the acid. The rate of release of phosphorus seems independent of the concentration of acetate added, but about 4 mg acetate is required for the release of 1 mg phosphorus, as can be seen in Figure 8-1, produced by Wentzel *et al.*(8).

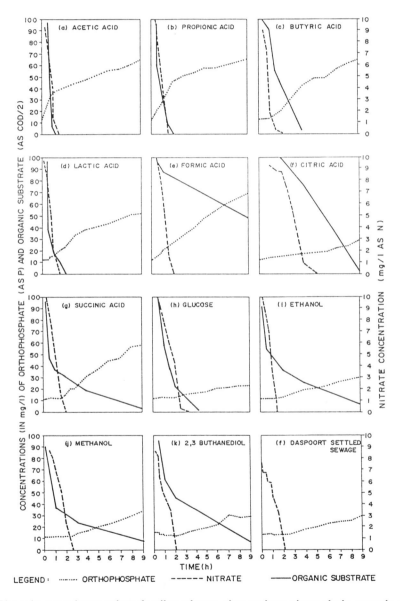

Figure 8-4. **Phosphorus release when feeding nitrate plus various short chain organic compounds.**

207

8.1.3 External Production of Volatile Fatty Acids

From the above discussions one can conclude that optimal biological phosphorus removal requires external generation of VFAs in order to limit the size of the contact zone and reduce secondary release of phosphorus. Early attempts at operating acid digesters for generating VFAs were frustrated by the onset of methane fermentation which put an abrupt halt to phosphorus removal in the activated sludge plant. Gerber(7) indicated that acetates are the most important VFAs. Since these are also the feedstock for methane-producing organisms, it follows that methane formation will reduce acetates drastically. In addition, analysis of the VFAs being produced in a digester is important. While measurements may indicate sufficient quantities of VFAs are present, they may not be in the right form to produce phosphorus release.

Barnard(5) suggested the use of primary sedimentation tanks (PSTs) and gravity thickeners as possible sources of acetates, since acid fermentation is initiated when primary sludge is accumulated in the conical bottoms of these tanks. Recycling of sludge elutriates the VFAs, allowing them to flow to the anaerobic or contact basin of an activated sludge plant. Recycling sludge to gravity thickeners produces a highly concentrated stream of VFA that can be directed to any section of a plant. This concept is referred to as "Activated Primaries" for the generation of VFAs(5). In the PST or thickener, the pH value should be maintained near neutral to ensure that the fermentation process produces mostly acetate and propionate. At lower pH values the end product of acid fermentation tends to be butyric acid. Elutriation of the acids will further stabilize the pH value. Unfortunately, recycling also favors the growth of the methane bacteria. After a few days of recycling the tanks may have to be cleared of sludge to prevent a build up of methane-producing organisms. Activated primaries operated along these lines have been successful in producing the necessary acetates for the proper functioning of Bardenpho plants or other variations of the same process.

8.2 Operating Experiences and Case Studies

Even though some of the original high rate plug flow plants that removed phosphorus are still being operated in that mode, little has been published about them lately. Studies, such as that of Milbury(1), indicated that it was possible to run these plants at high levels of reliability, even though at the time the mechanism was not clearly understood. In the process investigated, the influent side of the aeration train was aerated. Cutting down on aeration in this section, as described below, would have improved results.

Construction of nutrient removal plants started in late 1974. A number of plants were operated during the construction stage to remove phosphorus mainly by switching off some of the aerators at the inlet end of the aeration basin. At the Johannesburg Olifantsvlei plant, all but one of the aerators in the first zone of an extended aeration plant were switched off. It was possible to maintain the average effluent phosphorus concentration below 1 mg P/L for over a year. As the load to the plant increased, this became no longer possible.

A large number of plants have been constructed all over the world for the removal of phosphorus, in conjunction with nitrogen removal. At first, due to limited knowledge of the mechanism of the process, the anaerobic zones of the activated sludge plants were operated to produce the VFAs required for the growth of the phosphorus-removing organisms. Mixed results were achieved in meeting standards of 1 mg/L of phosphate measured as phosphorus, which is the most common requirement, even though in most cases removal of about 70% was achieved. The picture changed considerably after the introduction of externally generated VFAs or the addition of acetates to the anaerobic zone.

Examples are presented below of how the principles of biological nutrient removal are apparent in plant experiences. Additional summaries of plant operations are presented in the "Case Studies" section of Chapter 7.

8.2.1 Goudkoppies, Johannesburg

This was the first large scale plant to be designed in the Bardenpho mode. Plant statistics are given in Table 8-1. A flow diagram is shown in Figure 8-5. Each of the three modules was designed on the basis of treating a population equivalent to 250,000. At first the plant was designed for nitrogen removal only, with the idea of inducing phosphorus removal by making the third basin anaerobic. The anaerobic basin was added in late 1974 while the plant was already under construction. Its size was determined more on the basis of the space available than on process considerations.

The plant was to be served by two main sewers; one containing mostly domestic waste flowing directly from the central business district through a tunnel and the other a slow flowing main sewer delivering a mixture of domestic and stronger industrial wastewater at a level lower than the intake of the new plant. This latter sewer showed signs of decay and was to be replaced. Replacement of this sewer was postponed several times resulting in the plant being under loaded for the first few years of operation. Even today the sewer has not been replaced, but pumps have been installed to deliver its wastewater into the new plant.

Table 8-1. Plant statistics for the Goudkoppies plant(three modules each 50 ML/d capacity).

Primary Sedimentation	2 circular tanks each 29.5 m in diameter per module; Hydraulic surface loading at average daily wastewater flow: 1.5/h
Balancing Tank	Rectangular with baffles at base to produce channel conditions at low flow; Volume = 22,759 m^3; Nominal detention time at average daily flow = 3.6h (serves all three modules)

Biological Reactor

Compartments (in sequence)	Volume, m^3	Retention Time, h	Mechnical Equipment
Anaerobic	2,080	1	2 axial flow mixers, 38 rpm, 11 kW motors
Primary Anoxic	4,800	2.3	4 axial flow mixers, 30 rpm, 22.5 kW motors
Aerobic	14,700	7.1	12 aerators, 2,972 mm dia., 110 kW motors
Secondary Anoxic	4,800	2.3	4 paddle type stirrers, 7.3 rpm, 18.5 kW motors
Reaeration	2,700	1.3	2 aerators, 2,363 mm dia., 45 kW motors

Internal Recycling	One module; 4 spindle propeller pumps: 15 kW motors; Recycle rate = 4 times average daily wastewater flow per pump
Return Sludge	3 Archimedes screws per module: 1,200 diameter, 22.5 kW motors; Sludge recycle rate = 0.7:1 to 1.6:1 on average daily wastewater flow
Secondary Clarifier	4 tanks per module: 36.2 m diameter; Side wall depth: 2.4 m; Suction lift rotating bridge scrappers

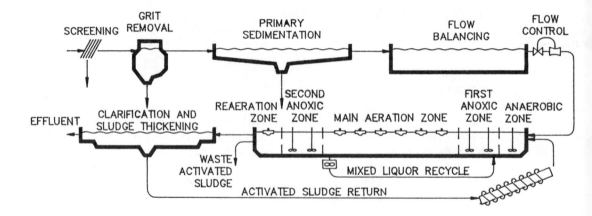

Figure 8-5. Flow diagram for the Goudkoppies plant.

When only the downtown sewer fed the plant, several operational shortcomings became evident.

a. In the under loaded condition, it was difficult to control the oxygen input to the plant. Severe bulking resulted from bad mixing and low oxygen tensions.

b. During weekends the exodus from the downtown area led to severe under loading which, combined with point (a) resulted in nitrate concentrations increasing over the weekends. This in turn upset the anaerobic basin, resulting in phosphorus concentrations rising on Tuesdays to about 2 mg P/L. This is illustrated in Figure 8-6. By Friday the effluent phosphorus would decline to as low as 0.1 mg P/L. The average could be maintained close to 1 mg P/L, but not below this mark.

c. The problem was further exacerbated by a number of drops in the feed channels which aerated the wastewater, while the screw pumps returning activated sludge to the anaerobic basin entrained much air.

d. The oxygen control system, consisting of probes activating an adjustable weir, was unreliable, making DO control very difficult, especially during storms which tended to wash higher loads of settled organic matter through the existing balancing tank.

The operation of the plant was greatly improved by installing pumps to lift the stronger industrial wastewater to the plant. The stronger wastewater fermented in the sewer producing sufficient VFAs to overcome the inherent problems of the plant, resulting in very reliable removal of phosphorus, as can be seen from Table 8-2.

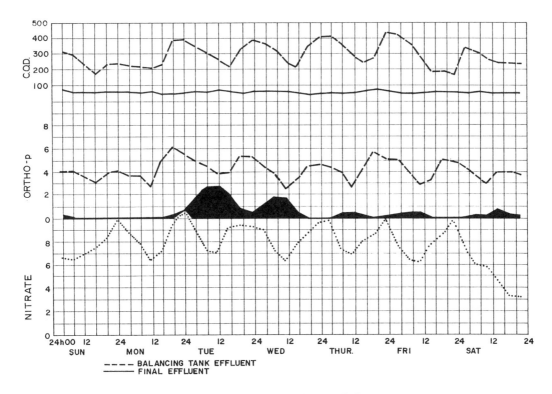

Figure 8-6. Initial Goudkoppies results showing "Tuesday Peaks".

Table 8-2. Goudkoppies results (mg/L) before and after contributions of septic outfall(4).

	No Septic Sewage Added				Septic Sewage Added			
Influent to reactor	COD	BOD	TKN	TP	COD	BOD	TKN	TP
Arithmetic mean	340	170	39	5.9	600	260	38	7.3
Effluent from reactor*	TP	OP-P	NH$_3$-N	NO$_3$-N	TP	OP-P	NH$_3$-N	NO$_3$-N
Arithmetic mean	1.11	0.89	1.33	4.51	0.66	0.36	0.39	1.60
Standard deviation	0.72	0.74	2.10	2.94	0.42	0.35	0.97	2.16
Geometric mean	0.93	0.48	0.13	2.91	0.58	0.27	0.01	0.54

*Approximately 1000 samples taken at 4 h intervals

211

8.2.2 Northern Works, Johannesburg

This plant is a virtual duplication of the Goudkoppies plant, but whereas the Goudkoppies plant could comply with the standards most of the time, even when under loaded, the Northern Works plant could not comply. The low COD:TKN ratio of less than 9:1 after primary sedimentation and the high phosphorus concentration of 16 mg P/L were partly to blame. COD:P ratios approached 40:1.

Attempts at acidifying sludge in an overloaded digester produced some success. However, after a short period of time methane-producing bacteria multiplied and removal deteriorated. Partial success was achieved when some of the primary sedimentation tanks were taken out of service, allowing fine suspended material to flow into the second anoxic zone which improved denitrification.

Substantial improvement in effluent quality was achieved when PST underflow was recycled to the head of the PSTs. However, effluent phosphorus deteriorated again and it became necessary to withdraw all the sludge from the PST from time to time to avoid methane development. Complete nitrate reduction remained a problem.

Finally, the anaerobic zone was split into four sequential reactors. The RAS was returned to the first reactor and the primary effluent was fed to the third. This removed the nitrates in the RAS to zero before the addition of the enriched PST effluent, leading to very good results. Some results covering this period are shown in Figure 8-7.

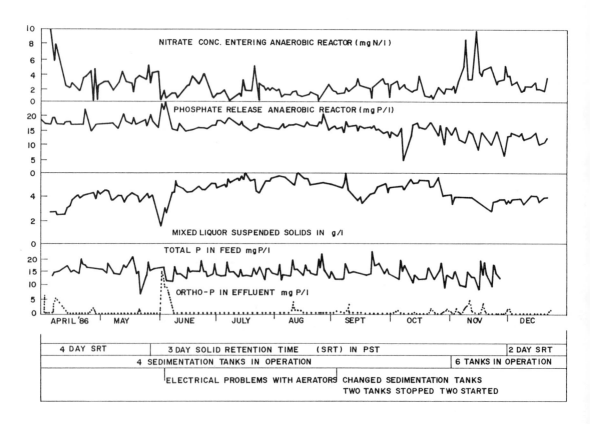

Figure 8-7. Results from Northern Works, Johannesburg.

212

8.2.3 Bulawayo, Zimbabwe

This plant was designed for a flow of 3 MGD(10 ML/d) of very strong domestic waste, having a BOD concentration of 500 mg/L, TKN of 70 mg N/L and total phosphorus of about 10 mg P/L for a population equivalent exceeding 100,000. A flow diagram is shown in Figure 8-8. Aeration was by means of surface aerators with draft tubes.

Since power was inexpensive and the waste sludge could be irrigated on dry lands, extended aeration was implemented. Tests showed that the peak BOD concentration during the day could be reduced by 50% through primary sedimentation. Two Dortmund type PSTs without mechanical equipment were provided for settling the waste by day and pumping the sludge to the plant at night, thus reducing the peak demand and avoiding installation of two additional aerators. Two pumps recirculated the primary sludge by day to prevent over thickening and blockages in the deep tanks and at night discharged the sludge to the activated sludge plant. Digesters could be added later to increase plant capacity.

Recycling of the underflow elutriated VFAs that formed during the day. Pumping the sludge out at night prevented the formation of methane. The plant has performed exceedingly well for about five years of operation, complying with the nitrogen standard of 10 mg N/L and the phosphorus standard of less than 1 mg P/L 95% of the time, with the average effluent phosphorus being less than 1 mg P/L.

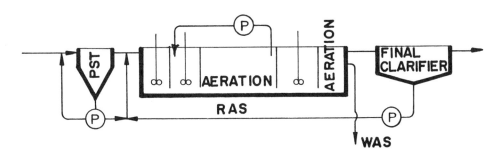

Figure 8-8. Flow diagram for Bulawayo, Zimbabwe.

8.2.4 Kelowna, British Columbia

The City of Kelowna is situated on pristine Okanagan Lake in central British Columbia. Waste characteristics are listed in Table 8-3. A flow diagram of the plant is shown in Figure 8-9.

Table 8-3. Raw waste characteristics for the Kelowna, British Columbia plant.

BOD_5	225 mg/L
Total Kjeldahl Nitrogen	30 mg/L
Total Phosphorus as P	7 mg/L
Suspended solids	200 mg/L
Mixed liquor temperature (est)	9 °C
Alkalinity(as $CaCO_3$)	200 mg/L
Flow	23 ML/d (6 MGD)

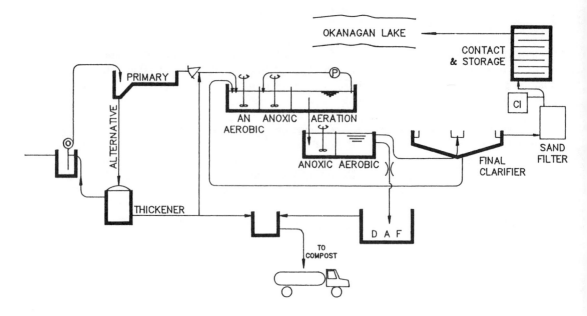

Figure 8-9. Flow diagram for Kelowna, B.C. plant.

The SRT needed for both nitrogen and phosphorus removal at low temperatures was calculated to be about 30 days. Each of the two modules of the plant consisted of 21 cells, arranged to allow maximum flexibility in changing the relative sizes of the various treatment zones.

The existing primary tanks as well as the thickener were incorporated into the new flow diagram. Since the plant was surrounded by houses, the sludge had to be trucked to a composting site. Secondary sludge is flotation thickened and mixed with the thickened primary sludge to minimize the volume for transport.

Careful consideration was given to phosphorus removal during the design stage. The wastewater is weak, winter temperatures are low and primary sedimentation was expected to selectively remove organic carbon. A number of steps were taken to address these adverse conditions. A by-pass was provided from the thickener underflow to the anaerobic basin, passing the sludge through a fine screen to extract the liquid. A control gate was provided at the effluent side of the PST to back up the flow and avoid aeration of the PST effluent. The plant had a modular construction for greater flexibility while ensuring that plug flow conditions are maintained. Some mixed liquor was by-passed from the head of the aeration section to the second anoxic zone to enhance denitrification.

The successful operation of the plant can be seen in Figures 8-10 and 8-11. The average total phosphorus in the first year of operation was 0.43 mg P/L. Nitrification was lost in the winter due to running the plant too close to the critical SRT. It is important to note that by increasing the SRT it was possible to regain nitrification during the cold spell. The mixed liquor temperature was estimated during design to drop to 9°C during the winter. During snow melts, the temperature dropped to 8°C for short periods without affecting plant performance.

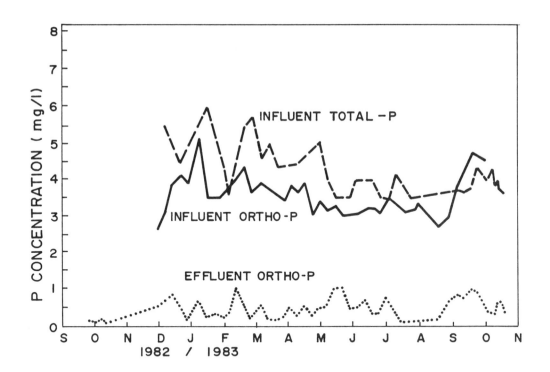

Figure 8-10. Kelowna plant effluent phosphorus.

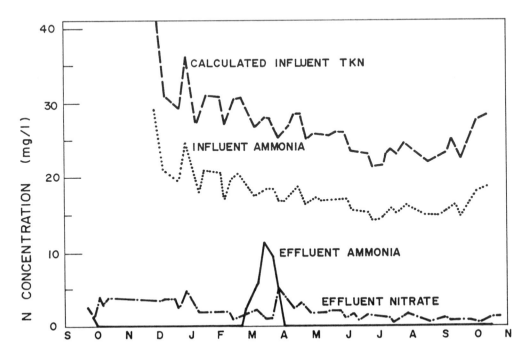

Figure 8-11. Kelowna plant effluent nitrogen.

215

The plant performed well without the by-pass from the underflow of the gravity thickener. This was attributed to the large volume of liquid passing through the gravity thickener, resulting from an operational requirement to pump sludge from the primaries until all the lines have been cleared of sludge. This had the effect of washing out some of the VFAs formed which were then returned to the PST and hence to the anaerobic zone. The VFA content of the settled sewage varied from 14 to 20 mg/L, giving a rough ratio of VFA to P of 4 to 1.

After about 15 months of operation, the by-pass from the thickener underflow was altered to direct the thickener supernatant directly to the anaerobic basins of the two modules. It was then possible to demonstrate the value of the VFAs by feeding the supernatant first to one module then to the other. The module receiving the supernatant removed phosphorus, the other lost the ability.

When attempts to optimize the thickener for VFA generation resulted in the loss of phosphorus removal through a cause still unknown, stirrers in the anaerobic basin were switched off. This resulted in a doubling of the release of phosphorus but in no overall phosphorus removal. This enhanced release of phosphorus was later surmised to be due to secondary release. The addition of sodium acetate to the anaerobic basin immediately restored phosphorus removal. Acetate formation in the thickener presumably failed either as a result of methane fermentation or the pH value of the thickener dropping too low.

Alum (20 mg/L) was added to the clarifier inlet resulting in effluent phosphorus concentrations of less than 0.1 mg P/L for up to three months at a time.

8.2.5 Secunda, Transvaal

The town serves a complex producing oil from coal situated in the catchment of the Vaal Dam which is the main source of water for about 6 million people. All effluents must comply with a phosphorus standard of 1 mg P/L and an ammonia standard of less than 10 mg N/L. The industrial effluent (which is larger than the domestic waste stream) is to be treated and reused resulting in a zero discharge. The first plant installed was a Bardenpho plant with combined nitrification and denitrification, as shown in Figure 8-12. Reasonable treatment results were obtained when it was under loaded, but eventually it could not remove more than 50% of the phosphorus.

In order to meet the standard, an industrial wastewater stream containing acetate was diverted to the plant with results shown in Figure 8-13. The short break indicates a period when no acetates were fed to the plant.

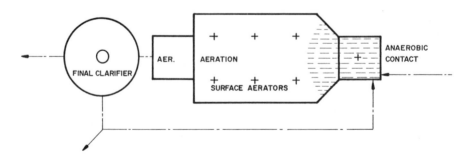

Figure 8-12. Secunda, Transvaal plant layout.

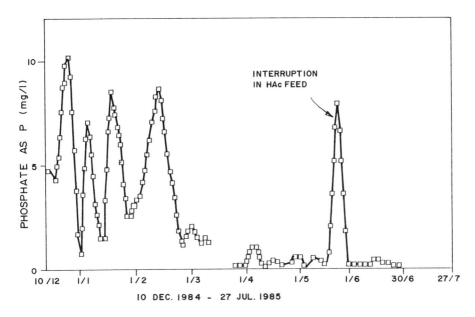

Figure 8-13. Secunda, Transvaal plant results after acetate addition.

8.2.6 Randfontein

About 25% of the load to this plant consists of effluent from the production of edible oil. The existing trickling filters reduced the BOD to 50 mg/L, probably due to the slow degradation of the oily wastes. The plant was expanded using a Bardenpho plant which receives 40% of the influent load while also treating the trickling filter effluent for nutrient removal. The flow diagram is shown in Figure 8-14.

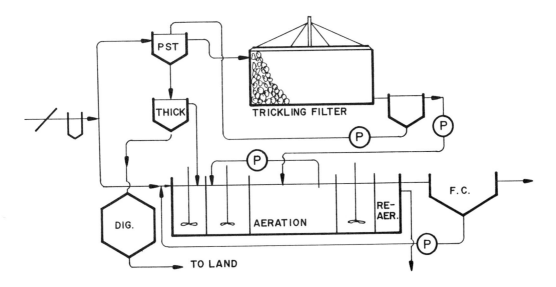

Figure 8-14. Randfontein plant flow diagram.

217

used in the aeration zones. Raw wastewater was fed directly to the activated sludge plant, but settled primary sludge from the trickling filter line was first passed to a thickener. Thickener supernatant was returned to the activated sludge plant. The trickling filter effluent was passed to the aeration zone. At the end of the aeration zone, the ammonia, nitrate and phosphorus in a filtrate of the mixed liquor were all near zero. This could only be explained by the slow degradability of the wastewater resulting in ample absorbed carbon being present in the lower level of the aeration basin, such that when the mixed liquor was recirculated through this zone complete denitrification took place.

The second anoxic zone received no nitrate and became anaerobic. With little available carbon remaining, the acetates generated there were insufficient to provide energy for the uptake of secondary-released phosphorus, resulting in the phosphorus profile shown in Figure 8-2. This profile is typical of three years of results. No amount of aeration after the anoxic zone could reduce the phosphorus to below 1 mg P/L. The second anoxic zone was then aerated and the effluent phosphorus concentration averaged 0.7 mg P/L.

8.2.7 Disney World, Florida

The Reedy Creek plant, serving the Disney World entertainment complex near Orlando, Florida, treats the equivalent waste of 300,000 persons per day. Even though designed in a high rate mode the plant normally produces a nitrified effluent. The plant configuration is the normal four pass system with aeration from spargers suspended from a walkway along one side of each pass. The effluent passes through a pond system and then through a wetland system. A dike was constructed to collect the effluent before passing it on to the receiving water body.

As can be seen from the results in Figure 8-15, little removal of phosphorus occurred in the wetland system. Eight of the aerators were switched off in the first pass in 1982 allowing some anaerobic conditions to develop. The phosphorus in the effluent subsequently was maintained below 0.6 mg P/L over a period of three years.

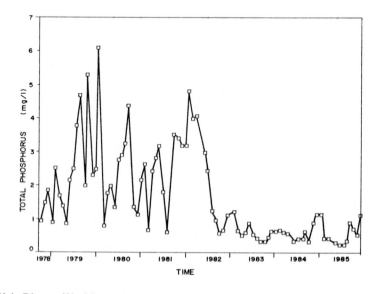

Figure 8-15. Walt Disney World treatment plant results for P removal.

8.2.8 Tembisa, Transvaal

The Tembisa plant expansion was designed to upgrade the existing trickling filter plant for biological nutrient removal. The trickling filter plant was designed for the treatment of 13 ML/d of high strength wastewater from domestic and industrial origin. The plant was to be expanded to treat 36 ML/d of wastewater having a COD concentration of just under 1000 mg/L giving the plant a population equivalent of just over 300,000 persons. Since the COD:TKN ratio was favorable at about 15:1, it was decided to expand the plant by adding a Three-stage Bardenpho process parallel to the existing plant while routing the trickling filter effluent through the new plant. A flow diagram is shown in Figure 8-16.

In the design, the trickling filter was loaded to hydraulic capacity (16 ML/d average daily wastewater flow) while 22 ML/d was diverted directly to the activated sludge plant. This wastewater flow was passed through an activated primary sedimentation tank (a-PST) for acid fermentation. The primary sludge was recycled through a fine screen to the inlet to the PST in order to elutriate the VFAs formed during the retention of the sludge in the PST. Recycling continued for one day, after which the sludge was sent to the digesters and the cycle repeated.

The overflow from the a-PST was fed into a plug-flow anaerobic basin after mixture with the WAS. Submerged stirrers kept the mixed liquor in suspension. The mixed liquor existing in the anaerobic basin was mixed with the recycled mixed liquor from the aeration basin and fed to the anoxic basin. The reactor was folded in the form of a "U" to facilitate the recycle of mixed liquor. One leg of the "U" was formed by the anaerobic basin, the anoxic basin and the first section of the aeration basin. The first aeration zone was furnished with four 75 kW surface aerators. The trickling filter effluent was discharged to this section of the aeration basin. The mixed liquor then proceeded to the second leg of the reactor which was furnished with six 55 kW aerators. Mixed liquor was recycled to the anoxic zone from a point between the second and third from the last aerators in the train. The effluent mixed liquor spilled over a manually adjustable weir to a distribution box and four 30 m diameter final clarifiers. Sludge was returned to the anaerobic basin by means of two centrifugal pumps per tank coupled directly to the underflow of each tank in pairs, thereby eliminating any possibility of air entrainment. Mixed liquor was recycled to the anoxic basin by using two submerged stirrers situated in front of openings in the wall on the floor of the aeration basin.

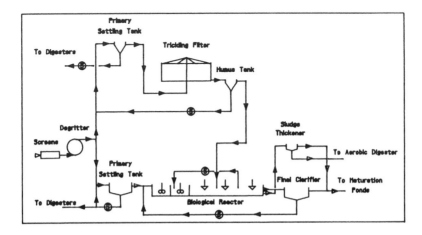

Figure 8-16. Flow diagram for Tembisa plant.

The performance record of dissolved oxygen (DO) meters in South Africa was so poor that it was decided to control the process by monitoring ammonia and nitrate in the effluent and taper the oxygen input accordingly. Manually adjustable overflow weirs controlled the immersion depth of the aerators. The aerators were programmed to start and stop by means of controls offered by a software package. After 24-h profiles of nitrate and ammonia in the effluent were determined, the program was set to control the times of operation of the various aerators. The adjustable weir was used to fine tune the oxygen input from day to day, depending on the results. Since the effluent standards required that the total nitrogen be below 10 mg/L while the effluent TP should be less than 1 mg/L, it was decided to err on the side of ammonia and to control the plant at less than 1 mg/L of nitrate-nitrogen and about 4 mg/L of ammonia-nitrogen.

Typical plant performance can be seen from average values presented in Table 8-4. The plant was operated at a SRT of 16 days which was controlled by wasting mixed liquor directly to a satellite clarifier from which all the underflow was wasted to digesters while the overflow was discharged to the effluent stream. Supernatant from the sludge digestion process was irrigated in a forest.

A model developed by the University of Cape Town (UCT) predicts that only the UCT process can produce reliable results and that the Three-stage Bardenpho process cannot be relied upon to remove phosphate due to an inability to remove nitrates(8). The model applied to the Tembisa plant predicts that the effluent nitrate-nitrogen concentration should be 13 mg N/L. If this were true, such nitrates would be recycled to the anaerobic basin with the RAS jeopardizing the conditioning of the sludge for phosphorus removal. Since phosphorus removal is excellent, this plant apparently is thus not behaving according to the model. The operation of other plants have been examined to shed some additional light on this dichotomy. The same model predicts that 7 mg/L of nitrate-nitrogen should be present in the effluent of the Kelowna plant with little phosphorus removal. This has been disproved by nitrate-nitrogen values consistently below 1 mg N/L and excellent phosphorus removal occurring at this plant.

The discrepancy between the model's predictions and field observations may be explained by two factors pertaining to the operation of this plant and others using point source aerators, i.e. aerators such as surface aerators, jet aerators and turbine aerators where the oxygen is introduced at a specific point while mechanical means are used to circulate the aerated mixed liquor to other points in the aeration basin. The first factor concerns the physical conditions in the basin where the rate of oxygen input and the rate of oxygen consumption are such that a gradient is formed across the aeration pocket. For example, consider a surface aerator with a draft tube. The mixed liquor must pass through a zone of high oxygen tension in order to be aerated, then it is forced down to the lowest level in the aeration pocket and again sucked into the draft tube. Efforts to keep the entire aeration basin aerobic will result in over aeration. If the design is such that the oxygen input is not sufficient to keep the entire basin aerobic, nitrates will be formed in the aerated section and denitrified in the under aerated section.

Table 8-4. Results of Tembisa plant (mg/L).

	Influent	Effluent
COD	960	40
BOD	510	4
TKN	65	<1
Nitrate-N	--	1.5
Ammonia-N	--	4.3
Total P	12	0.8

The second factor playing a role is the observed storage of glycogen in bacteria when they are exposed to a high concentration of VFAs. Alleman *et al.*(9) observed this phenomenon in sequential batch reactors, while Ekama *et al.*(10) referred to this as the selector effect. The generation of VFAs in the a-PST allows the exposure of the RAS to a high concentration of VFAs in the mixing zone of the plug-flow anaerobic basin. This seems to result in a reservoir of stored COD which will last well into the aeration zone. This is demonstrated in Figure 8-17 which shows measured oxygen uptakes rates (OUR) as opposed to those predicted by the UCT model. The presence of such a store of COD would lead to a high degree of denitrification taking place in the aeration zone when anoxic zones are formed. This was not taken into account in the available design models, but was considered in the design of the Tembisa plant. The design of the Tembisa plant was based on experience in previous plants indicating that up to 40% of the nitrogen removal could take place in the aeration basin, allowing the use of smaller formal anoxic zones or deleting the second anoxic zone in the Five-stage Bardenpho process while still removing virtually all the nitrate formed.

The percentages given in Figure 8-18 show the nitrogen loss in the various sections of the reactor during the period of testing. The relatively low loss in the formal anoxic basin is in part due to the low mass recycle of nitrate to this unit resulting from the high rate of removal in the aeration basin. There is also a remarkable loss of nitrate in the trickling filter. This may have resulted from high rate recycling of humus tank underflow to the PSTs.

The COD:TKN ratio was reduced to less than 10:1 by the reduction of carbonaceous compounds in the trickling filter. Even so, it would appear that the capacity for the system to remove nitrogen has not been reached. Nitrate reduction remains an important issue when considering phosphorus removal. Since this plant succeeds in removing virtually all the nitrate, little difficulty is experienced in also removing phosphate.

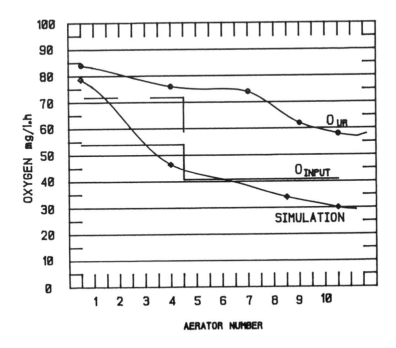

Figure 8-17. Oxygen uptake rates through Tembisa plant.

221

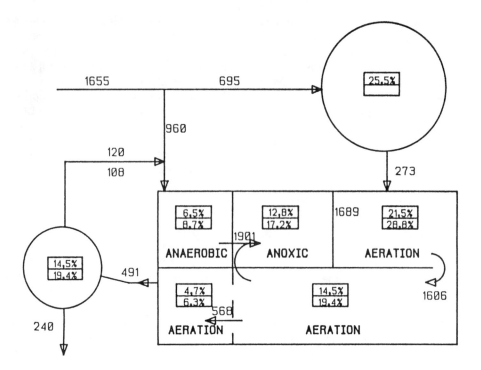

Figure 8-18. Nitrogen mass balance through Tembisa plant.

The UCT model based phosphorus removal on the ability of the plant to remove nitrate-nitrogen to quite low levels. The reason for the great discrepancy between predicted and actual values can be traced back to the laboratory base of this model. If the plants were in all respects similar to the laboratory units, the model might have applied. However, plants with point source aerators differ considerably from fully aerated, fully mixed plants. Even when using fine bubble diffused air, the results tend to be much better than those predicted by the model. Even here it is suspected that some simultaneous nitrification and denitrification is taking place.

In the Kelowna example discussed earlier, the UCT model predicted effluent nitrate-nitrogen concentrations of about 7 mg N/L, while the plant is consistently achieving less than 1 mg N/L. The Orange County plant in North Carolina is achieving more than 80% nitrogen reduction and an effluent phosphate concentration of below 1 mg P/L without any formal anoxic zones. Point source jet aerators separate the mixing from the aeration function. Simply by regulating the air input, almost any degree of nitrogen removal is possible.

Phosphate removal in the Tembisa plant is very dependent upon the VFA generator, in this case an activated PST. Figure 8-19 shows very clearly the effect of bringing the tank on-line and taking the tank off-line. It is very difficult to quantify the VFA generation in the PST and it is not routinely done for plant operation.

The plant receives industrial waste which is the cause of wide fluctuations in the load. Timer switches on the aerators allow the operator to control the plant within the requirements of less than 1 mg P/L on average. Occasional effluent phosphorus concentrations of just more than 1 mg P/L and one value above 2 mg P/L in a six month period could be reduced with improved controls.

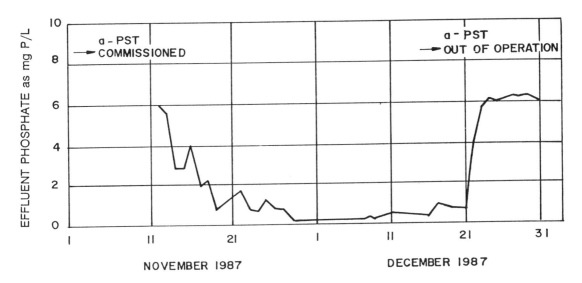

Figure 8-19. Effect of a-PST on phosphate removal.

8.2.9 Summary of Plant Experiences

These case studies were selected to give the reader an idea of the problems that were experienced in the development of the process. Some of the earlier plant experiences pre-dated expected phosphorus standards, which allowed some time for experimentation. The only problems experienced with plants that have come on-line recently relates to mechanical or computer problems. The use of computers for plant control in areas of high lightning intensity is risky and as yet is not proven. Another problem relates to toxicity. Little is known about the effect of toxics on the behavior of the phosphorus-removing organisms.

8.3 Special Considerations in Operating for Phosphorus Removal

8.3.1 Sludge Age or SRT Control

Apart from DO control, SRT is the most important control parameter for phosphorus removal. In a high rate system the SRT must be controlled to be above about 4 days, but low enough to avoid nitrification. In warmer climates, nitrification may be unavoidable and a pre-anoxic system should be used for preventing nitrate from interfering with phosphorus removal. Fortunately, in warmer climates, sludge is more active and very rapid denitrification normally takes place.

In a combined system for removing both phosphorus and nitrogen, the SRT must be sufficient to allow for nitrification at all times, taking into account the anoxic and anaerobic zones. The SRT of the latter is determined by the relevant rates of denitrification and the choice of a safety factor.

Volumetric control is the simplest way of ensuring adherence to prescribed SRTs. It lends itself to computer control for more sophisticated plants. Since the most desirable way of reducing the water content of sludge is flotation thickening, wasting mixed liquor instead of clarifier underflow has many advantages and few disadvantages.

223

8.3.2 Dissolved Oxygen (DO) Control

Some DO control may be required in high rate plants, mainly to ensure that there is sufficient DO for uptake in the aeration basin and to avoid subsequent release in the final clarifier. Phosphorus removal in combined systems is dependent first and foremost on good DO control. Many DO control systems have been abandoned making it imperative that the selection of one be done carefully to find one which is reliable. The efficiency of phosphorus removal in many full-scale plants has been found to be totally related to the efficiency of the DO control system. Thus back-up systems are essential.

There are two options for control of surface aeration systems; the first being an adjustable weir controlling the immersion depth of the aerators and the second being the capability to switch aerators on and off in a preselected sequence depending on the demand for more or less air from the metering system. Timer switches may then serve as a back-up DO control system. When using adjustable weir control, allow for a long stationary weir at the level of maximum immersion to avoid tripping out of the aerators during storm flow conditions, which could lead to the release of phosphorus in the aeration basin. It may also be necessary to provide for storm by-passes directly to the clarifiers to protect the aerators from tripping.

Control of oxygen in bubble aeration plants is more difficult, especially when there are a number of modules feeding from a central blower house. With every change in blower output, the flow of air to every zone of every module varies and needs to be readjusted. One approach for overcoming this limitation is to use an average of the output of probes placed in all sections of a plant for the control of the air output from the blowers. Air flow meters to all the modules could then be used in a separate control loop to ensure that the air distribution to the modules will always be in a preset proportion. A second independent control loop could then be used to ensure that the air flow to each of the zones in the aeration basin is distributed in a preset ratio to ensure tapering of the DO towards the end of the aeration basin. This reduces the reliance on DO meters for all the controls.

A second method of control is the use of a pressure switch on the main delivery line to ensure that the pressure in the main feeder is maintained at a preset value at all times. DO meters in each zone of each module could then open valves on demand to maintain the DO concentration in that zone at a preset value. The disadvantage of this system is that DO meters are not that reliable and, whereas an operator would rely on the average value of a number of outputs in the first alternative, an operator is totally reliant on every DO probe in every section of the plant when using the second alternative. If one meter is defective and registers zero, that valve will open fully and over aerate that section, leading to nitrate appearing in the effluent, while under aerating the other modules or bringing in too much power.

Back-up systems for a central blower system may consist of a programmed oxygen demand, simulating the diurnal variation in DO demand for the plant.

8.3.3 Scum Control

Nutrient removal plants tend to encourage scum formation. All of the treatment zones should be designed for moving scum forward to the final clarifiers where removal is essential. The scum cannot be scraped up a beach and final clarifiers should have a system of positive withdrawal of scum. Any recycle of the scum would lead to rapid growth and severe problems. The scum should be diverted to dissolved air flotation thickening tanks for final disposal.

8.3.4 Level of Operator Skill

The level of operator skill required will depend largely on the sophistication of the plant. Smaller plants could be designed with less sophistication. However, operators should be well versed in plant management and preventative maintenance. For example, while loss of power for a period up to an hour or two may not be serious in conventional activated sludge plants, such outages may result in the release of some of the phosphorus accumulated in the sludge, thereby exceeding effluent limits. If power outages cannot be avoided, standby chemicals may be required to correct for such mishaps.

Preventative maintenance is essential for preventing plant breakdowns, especially of those mechanical components that play an essential role in nitrate and phosphorus removal such as stirrers, recycle pumps and control equipment.

While highly skilled operators may not be essential for the operation of nutrient removal plants, a disinterested operator can spell disaster. Many plants are being operated at present by lesser skilled operators taking a strong interest in the results produced, thereby producing effluents of excellent quality.

8.3.5 Control of Nitrogen

Nitrification in high rate systems should be limited by controlling the SRT in order to avoid interference with phosphorus removal.

Some operators control aeration by monitoring the ammonia and nitrate in the effluent. Periodic diurnal profiles of nitrates and ammonia may show periods of under or over aeration and give the operator some pattern for setting timer switches on the aerators. Once the pattern has been set, analysis of a composite sample serves to confirm the setting.

Special kits with made up chemicals enable even lesser equipped plants to determine effluent ammonia, nitrate and phosphorus concentrations. Experience shows that operator interest and involvement in the performance of the plant is improved considerably by providing daily information or allowing the operator to perform simple tests.

8.3.6 Redox Control of the Anaerobic Basin

Since the emphasis on acetate production has shifted to the activated primary sedimentation tanks, interest in redox control of the anaerobic zone has waned. It is now considered counter productive to enclose the anaerobic zone or to allow the redox to drop to low values. There would appear to be an advantage in controlling the redox potential above a certain value to avoid secondary release of phosphorus in anaerobic or anoxic zones. This may be done by using spargers in the contact zone or purposely allowing some nitrates to be returned.

8.3.7 Control and Monitoring of the Acid Generator

It unfortunately is difficult to measure acetates as a routine test unless equipped with a gas chromatograph. Even then the determination of VFAs at low concentrations is not very reliable when using standard methods. Since few laboratories have the equipment to perform acetate analyses, any new plant must initially have some analyses done at a reputable laboratory to establish the best operational procedure for that particular plant.

Recycling of primary sludge in the activated primary is useful for elutriating the acetates formed but may encourage the growth of methane bacteria. When these take over, the activated sludge plant immediately looses ability to remove phosphorus. The causes have not been established beyond doubt, but through experience it has been found that at about 20°C all the sludge in the PST must be removed after 3 to 4 days of recycling. With more than one tank available, a program of recycle and draw-off can be established to wash out methane formers. Recycle lines for the two tanks must be kept separate to avoid any inoculation of "new sludges". The problem of methane formation seems to be more pronounced in the warmer climates, while in cooler climates it appears more probable that the process of acid formation could be continuous. The design should allow for some flexibility in order to operate the plant optimally.

8.3.8 Chemical Back-up Requirements

It is impossible to guarantee a phosphorus standard while relying on biological removal only, even though some plants have averaged lower than 1 mg P/L over a full year of operation. Mechanical failures of essential equipment, such as stirrers, aerators or pumps may cause excessive release of phosphorus. Toxic discharges also may upset the mechanism of phosphorus removal.

A chemical back-up system could serve as an incentive to optimize biological removal at a plant and to have a good preventative maintenance program. The annual consumption of chemicals acts as a indication of biological treatment performance. It could also serve as a justification for additional standby equipment or improved control strategies. Preliminary evidence is that chemical addition complements biological removal, i.e. the addition of small doses of chemicals improves rather than hinders the biological removal mechanism.

Magnesium and potassium take part in the biological removal of phosphorus. In wastewaters where these elements are in short supply, they may need to be added. When failure to remove phosphorus occurs under otherwise favorable conditions, it may do well to check for a shortage of either of these elements.

8.3.9 High Level Phosphorus Removal

Most effluent phosphorus standards require the effluent total phosphorus to be less than 1 mg P/L. A number of full-scale plants produce a filtered effluent of less than 0.1 mg P/L while the total phosphorus of the unfiltered effluent would average about 0.5 mg P/L. It would be difficult to achieve standards of less than 1 mg P/L without filtration or some form of tertiary treatment such as chemical treatment and filtration. At the Kelowna, B.C. plant it has been possible to reduce the effluent phosphorus from about 1 mg P/L to less than 0.1 mg P/L by the addition of about 20 mg/L of alum to the clarifier feed before filtration. At Vereeniging near Johannesburg, a standard of 0.15 mg P/L is maintained by a combination of biological treatment, chemical precipitation and sand filtration.

Biological phosphorus removal could, therefore, still be part of a treatment scheme for the removal of effluent phosphorus to much lower levels. Some experimentation would be required for design of the plant.

8.4 Cost Implications

The greatest folly that can be perpetrated in comparing costs for nutrient removal is to generalize or to use generalized information. Each individual case must be considered on its merits, taking into account the local conditions, local costs, the particular characteristics of the sewage, the general mean temperatures, the influx of groundwater, the nature of groundwater, the cost of alternative methods, the proposed nature of sludge treatment and many more conditions.

The Reedy Creek plant is a case in point. Note from Figure 8-15 the effect of switching off some air at the upstream end of the aeration basin. Not only did the effluent nitrogen and phosphorus improve, but the suspended solids over the last 18 months of plant operation dropped to consistently less than 5 mg/L. With these improvements, the effluent BOD before wetland treatment also dropped to less than 5 mg/L. It is not clear if the overall improvement in effluent quality came at a savings in power. Most probably the air requirements for the remainder of the aeration basin increased so as to off-set the savings from switching out aerators. However, the increased nitrogen removal would indicate that some nitrate was denitrified which could be interpreted as an overall savings in power. It would appear that, in general, a savings of power brought about an improvement in effluent quality to which one may attribute a cost benefit, depending on the conditions. If flotation was the selected mode of sludge thickening, there would probably be a little increase in cost due to this special treatment of the sludge.

In this case the probable reason for the excellent results at little cost was the site conditions. The high ambient temperatures and the flatness of the terrain must have resulted in a fair degree of acid fermentation taking place in pump sumps, force mains and holding tanks. All that was needed was a contact zone for triggering the process.

In many plants denitrification can be brought about by purposely under aerating which may result in a savings in costs accompanied by an improvement in effluent quality. Needless to say, if the plant is already over loaded, this may not be possible.

When designing a new plant for phosphorus removal only, one may have a situation where the VFAs are abundant in the influent. Provision must then be made for a contact zone in the aeration basin, while flotation thickening of sludge must be the choice. The latter may be the most cost effective way of dealing with the sludge, while the cost of the anaerobic contact zone should be minimal. The cost of the aeration basin structure may vary between 8 and 15% of the total plant cost. Adding a contact zone may come to as little as 1% of plant cost. If the influent stream contains little VFAs, a few additional pumps would be required for recycling sludge in the PST. Again the cost will be low compared with the total plant cost.

When designing a plant for both nitrogen and phosphorus removal at all times of the year, the plant will be more costly than a high rate plant, but the picture must always be seen in perspective. The cost of biological nitrogen removal must be compared with alternative methods of nitrogen removal. If biological removal is more cost effective, the additional cost for phosphorus removal is small, as discussed above. Except where space is a problem, a degree of internal denitrification is cost effective where nitrification is a primary requirement.

Nitrogen removal in cold climates at all times of the year is costly and the need must be clearly established. However, in Denmark and the Netherlands, with mildly cold climates, biological nitrogen removal is seen as cost effective. Under these conditions, one may find that biological removal of phosphorus will also be cost effective.

8.5 References

1. Milbury, W. F., D. McCauley and C. H. Hawthorne. Operation of conventional activated sludge for maximum phosphorus removal. **Jour. Water Pollut. Control Fed., 43**(9), 1890, 1971.

2. Barnard, J. L. Cut P and N without chemicals. **Water and Wastes Eng., 11**(7), 33, 1974.

3. Fuhs, G. W. and M. Chen. Microbiological basis of phosphate removal in the activated sludge process for the treatment of wastewater. **Microbial Ecology, 2**, 119, 1975.

4. Nicholls, H. A. and D. W. Osborn. Bacterial Stress: A prerequisite for biological removal of phosphorus. **Jour. Water Pollut. Control Fed., 51**(3), 557, 1979.

5. Barnard, J. L. Activated primary tanks for phosphate removal. **Water S.A., 10**(3), July 1984.

6. Wells, W. N. Differences in phosphate uptake rates exhibited by activated sludges. **Jour. Water Pollut. Control Fed., 41**(5), 765, 1969.

7. Gerber, A., E. S. Mostert, C. T. Winter and R. H. de Villiers. The effect of acetate and other short chain carbon compounds on the kinetics of biological phosphorus removal. **Biennial Conference, Southern African Branch of Inst. of Wat. Poll. Contr.**, May 1985.

8. Wentzel, M. C., P. L. Dold, G. A. Ekama, and G. v. R. Marais. Kinetics of biological phosphorus release. **IAWPRC Post Conference Seminar**, Paris, September 1984.

9. Alleman, J. E., and R. L. Irvine. Storage-induced denitrification using sequencing batch reactor operation. **Water Research, 14**, 1488, 1980.

10. Ekama, G. A., G. v. R. Marais and I. P. Siebritz. Biological excess phosphorus removal. Chapter 7. **Theory, design and operation of nutrient removal activated sludge processes.** Water Research Commission, P. O. Box 824, Pretoria 0001, 1984.

Index

FRP. *See* Fiberglass reinforced plastic
Full radius "ducking skimmer/rotating
    weir" scum removal systems, 61
Full-scale experience
  in biological nitrogen removal, 71–88
  in biological phosphorus removal, 159, 188–
    200, 203, 208–223
  in chemical phosphorus removal, 130–139

Gas-filled packed bed systems, 46
Gas/liquid ratio, 37
Glucose, 147
Goudkoppies, Johannesburg plant, 203, 209–
    211
Grating, 125, 126
Great Lakes Drainage Basin, 1

Hampton Roads Sanitation District
    (HRSD), Virginia, 79–81, 152, 173–174,
    177, 194–198
Handling
  of liquid chemicals, 119
  of methanol, 52
  of scum, 61
  of sludge, 127–129, 186
  of solids, 129
Henry's Law, 37
Heterotrophs, 18, 22, 30
High-porosity systems, 46
Hookers Point Wastewater Treatment Plant,
    Tampa, Florida, 72–73
HRSD. *See* Hampton Roads
    Sanitation District
Hydrated lime, 70
Hydraulic loading, 39
Hydraulic residence time, 186
Hyperion plant, Los Angeles, California, 142

IAWPRC model, 55
Incineration, 128
Inhibitory compounds, 14
Intracellular storage, 148
Ion exchange, 34, 40
Iron salts, 94, 96–104, 106, 107, 111, 112, 128.
    See also specific types

Jar tests, 117
Jones Island Wastewater Treatment Plant,
    Milwaukee, Wisconsin, 131–132

Kelowna, British Columbia plant, 206, 213–216
Kinetics of biological nitrogen removal, 43
Kjeldahl nitrogen, total (TKN), 14, 55, 56, 62,
    152

Lactic acid, 147
Lambers Point Wastewater Treatment Plant,
    79–81
Landfilling, 129
Landis Sewage Authority Wastewater Treat-
    ment Plant, Vineland, New Jersey, 81–82
Largo Wastewater Treatment Plant, Largo,
    Florida, 76–78, 192–193
Laundry detergents, 92
Lime, 8, 40, 107
  alkalinity demand for, 95
  in biological nitrogen removal, 35, 39, 70
  in biological phosphorus removal, 187
  in chemical phosphorus removal, 94–96, 111–
    113
  costs of, 187
  feed system for, 178, 183
  hydrated, 70
  pH increase and, 38
  precipitation of, 96
  solubility curve for, 95
Lipids, 147. *See also* specific types
Liquid chemicals, 119, 120. *See also* specific
    types
Liquid-filled packed bed systems, 46
Looped reactors, 49
Lower Potomac Water Pollution Control Plant,
    Fairfax County, Virginia, 135–136
Lower Susquehanna River Basin, 1
Low-head submersible non-clog sewage
    pumps, 60
Low-porosity packed bed systems, 46
Low-speed aerators, 48
"Luxury Uptake" of phosphorus, 142
Lysis, 5, 47

Magnesium, 147
Magnesium hydroxide, 40
Mainstream processes in biological phosphorus
    removal, 179–180, 184, 188
Maintenance, 49, 51–52, 116
Mass balance checks, 57
MCRT. *See* Mean cell residence time
Mean cell residence time (MCRT), 46, 54

Quality requirements for effluent, 34
Quicklime, 70, 187. *See also* Lime

Randfontein plant, 217–218
RAS. *See* Return activated sludge
RBC. *See* Rotating biological contactors
Reactors. *See also* specific types
  anoxic, 47
  batch, 48, 71, 85–86, 147, 151
  design of, 178
  endless loop, 48
  fixed film, 18, 29–33
  looped, 49
  overflow from, 183
  packed bed, 29, 46, 47
  sequencing batch, 48, 71, 85–86, 151
  submerged packed bed, 29
Reagents, 40. *See also* specific types
Recarbonation, 39
Recirculation, 30
Recycle pumping, 60, 64–65, 69
Redox control of anaerobic basin, 225
Reedy Creek plant, 227
Regenerant recovery, 40
Reno-Sparks Wastewater Treatment Facility,
    Cities of Reno and Sparks, Nevada, 73–
    74, 190–191
Reoxidation, 146
Residence time
  anoxic, 26, 27
  hydraulic, 186
  mean cell (MCRT), 46, 54
  overall, 27
  required, 27
Respiration, 19, 20
Retention time
  anoxic, 67
  solids. *See* Solids retention time (SRT)
Return activated sludge (RAS), 51, 167, 169,
    171, 175, 182, 183
Rillings Road plant, San Antonio, Texas, 142
River Oaks Advanced Wastewater Treatment
    Plant, Hillsborough County, Florida, 75–76
Rotating biological contactors (RBC), 29, 33–34

Safety, 11, 52, 54, 117, 119, 125
Sand, 46
Saturation coefficients, 9
SBOD. *See* Soluble biochemical oxygen
    demand

SBR. *See* Sequencing batch reactors
Scum
  control of, 224
  organisms for production of, 67
  removal of, 61, 67
Seasonal nitrification, 65
Secondary clarification, 57, 61, 66–67, 122, 127
Secondary sludge production, 55
Secunda, Transvaal plant, 216
Sedimentation, 92, 93, 208
Selective ion exchange, 34, 40
Seneca Falls plant, 147
Separate stage system. *See* Two-sludge
    system
Sequencing batch reactors, 48, 71, 85–86, 151
Shipping costs, 187
Short-chain carbohydrates, 206–207. *See also*
    specific types
Short-chain fatty acids, 145. *See also* specific
    types
Sidestream process. *See* Phostrip process
Simultaneous precipitation, 94
Single-sludge systems, 22–25, 34, 43–45,
    47–49. *See also* specific plants
    by name
  advantages of, 47, 71
  aeration systems in, 63–64
  basins in, 47, 48, 62–63
  capital costs of, 62
  costs of, 49, 52, 53, 62, 63, 65, 67, 69
  design of, 53–65
  facilities costs in, 62, 65
  facilities design for, 57
  maintenance of, 49, 51–52
  mass balance checks in, 57
  mixers in, 64
  operating costs of, 69
  operation of, 51–52
  operational costs of, 67
  performance of, 49, 51, 86
  process design in, 53–57
  recycle pumping in, 64–65
  secondary clarification in, 61
  stoichiometry of, 49, 51
  summary of, 53
Slaking process, 70
Sludge. *See also* Solids
  activated. *See* Activated sludge
  age of, 9, 11, 13, 223
  bulking, 66, 67